Digital Signal Processing

A Primer With MATLAB®

Digital Signal Processing

A Primer With MATLAB®

Samir I. Abood

CRC Press
Taylor & Francis Group
Boca Raton London New York

CRC Press is an imprint of the
Taylor & Francis Group, an **informa** business

MATLAB® is a trademark of The MathWorks, Inc. and is used with permission. The MathWorks does not warrant the accuracy of the text or exercises in this book. This book's use or discussion of MATLAB® software or related products does not constitute endorsement or sponsorship by The MathWorks of a particular pedagogical approach or particular use of the MATLAB® software.

CRC Press
Taylor & Francis Group
6000 Broken Sound Parkway NW, Suite 300
Boca Raton, FL 33487-2742

First issued in paperback 2022

© 2020 by Taylor & Francis Group, LLC
CRC Press is an imprint of Taylor & Francis Group, an Informa business

No claim to original U.S. Government works

ISBN-13: 978-0-367-44493-8 (hbk)
ISBN-13: 978-1-03-233716-6 (pbk)
DOI: 10.1201/9781003010548

Publisher's Note

The publisher has gone to great lengths to ensure the quality of this reprint but points out that some imperfections in the original copies may be apparent.

Visit the Taylor & Francis Web site at
http://www.taylorandfrancis.com

and the CRC Press Web site at
http://www.crcpress.com

Dedicated to

My great parents, who never stop giving of themselves in countless ways,
My beloved brothers and sisters;
My dearest wife, who offered me unconditional love
with the light of hope and support;
My beloved kids: Daniah, and Mustafa, whom I can't force myself to stop loving;
To all my family, the symbol of love and giving.

Contents

Preface

Digital signal processing (DSP) denotes various techniques for improving the accuracy and reliability of digital communications. The philosophy behind DSP is quite complicated. Digital signal processing converts signals from an analog form into digital data that can then be analyzed and consequently turned back into an analog signal with improved quality after the DSP system has finished its work.

In DSP, the engineers usually study digital signals in one of the following domains: time domains, frequency domains, spatial domains, and wavelet domains.

The applications of DSP include digital image processing, audio signal processing, audio compression, speech processing, video compression, digital communications, digital synthesizers, speech recognition radar systems, ultrasound and sonar, financial signal processing, seismology, and biomedicine.

The DSP algorithms can be run on general-purpose computers and implemented using software code program, and Simulink also can implement by using hardware as modern technologies for digital signal processing include more powerful general-purpose controllers, microprocessors, stream processors, and field-programmable gate arrays (FPGAs).

The typical processing approach in the time domain is an improvement of the input signal through a method called filtering. In digital filters there are two types of filter with and without feedback. There are various ways to characterize filters: as a linear filter, causal filter, time-invariant filter, stable filter, finite impulse response (FIR) filter, and infinite impulse response (IIR) filter.

The signals are converted from the time domain to the frequency domain usually through the use of the Fourier Transform. The Fourier Transform, also called spectrum or spectral analysis, converts the time information into a magnitude and phase component of each frequency. The engineer needs to study the spectrum to control which frequencies are present in the input signal and which are missing.

Digital filters originate in both IIR and FIR types. While FIR filters are always stable, IIR filters have feedback loops that may become unstable and oscillate. Digital filters can be analyzed through the z-transform, which provides a tool for analyzing stability issues of digital IIR filters. Also, it is analogous to the analyse and designs analog IIR filters represent the Laplace transform

The book offers a good understanding of a signal's behavior and its applications. The book begins with the study of signals and systems. Then it presents their applications in the different types of configurations shown in lucid detail. The book presents the relation of signals and systems. This book is intended for college students, both in community colleges and universities.

This book is organized into 16 chapters. With a short review of the basic concept of continuous and discrete signals in Chapter 1, it starts with a discussion of continuous and discrete signals and the generation of continuous and discrete signals in MATLAB®. It also discusses the classification of signals and systems, MATLAB in DSP, and the applications of DSP. Chapter 2 presents signal properties as periodic and aperiodic sequences, even and odd parts of a signal, transformations of the independent variable, and linear time-invariant causal systems (LTI). The description of linear convolution, convolution properties, and types of convolutions is discussed in Chapter 3.

Chapter 4 covers difference equations, and system representation using impulse response. Chapter 5 deals with the Discrete-Time Fourier Series (DTFS) coefficients of periodic discrete signals. It includes the Discreet Fourier Series of the discrete systems. Discrete-Time Fourier Transform (DTFT), frequency response, DTFT for discrete signals, and the interconnection of systems in the frequency domain are covered in Chapter 6. Discrete Fourier Transform algorithms (DFT) are included in Chapter 7. The chapter also elaborates on the method of decimation-in-frequency and in time.

In Chapter 8, the principles of the Fast Fourier Transform, decimation-in-frequency method, and decimation-in-time method are examined. Chapter 9 discusses the z-transform, region of convergence (roc), properties of the z-transform, and it also discusses the inverse of z-transform. Chapter 10 describes the z-transform applications in DSP for evaluating LTI system responses using z-transform and implementation of the system using z-transform. Chapter 11 introduces pole-zero stability, difference equations and transfer function, and the stability of DSP systems.

Chapter 12 discusses sampling relating the FT to the DTFT for discrete-time signals, the sampling of continuous-time signals, and instantaneous sampling. The description of digital filters and filter types and specifications is discussed in Chapter 13. Chapter 14 presents the implementation of IIR digital filters and their properties and the design of a notch filter by MATLAB.

Chapter 15 deals with the implementation of Finite Impulse Response (FIR), and it is design.

Chapter 16 deals with the digital filter design, the realization of digital filters, and direction-form I realization.

Earlier experience using the MATLAB program is not needed since the author highly recommends that the reader studies this material in conjunction with the MATLAB Student Version. Chapter 1 and Appendix C of this text provides a practical introduction to MATLAB.

MATLAB® is a registered trademark of The MathWorks, Inc. For product information, please contact:

The MathWorks, Inc.
3 Apple Hill Drive
Natick, MA 01760-2098 USA
Tel: 508 647 7000
Fax: 508-647-7001
E-mail: info@mathworks.com
Web: www.mathworks.com

Acknowledgments

I appreciate the suggestions and comments from several reviewers, including assistance from Prof. Zainab Ibrahim/University of Baghdad/Electrical Engineering Department, Dr. Muna Fayyadh/ Colorado Technical University, and special thanks to Nafisa Islam/Prairie View A&M University Their frank and positive criticisms led to considerable improvement of this work.

Finally, I express my profound gratitude to my wife and children, without whose cooperation this project would have been challenging if not impossible. We appreciate feedback from students, professors, and other users of this book. I can be reached at sameeralrifaee74@ieee.org sabood@ student.pvamu.edu and sameeralrifaee74@gmail.com.

Author

Samir I. Abood received his BSc and MSc from the University of Technology, Baghdad, Iraq, in 1996 and 2001, respectively. From 1997 to 2001, he worked as an engineer at the same university. From 2001 to 2003, he was an assistant professor at the University of Baghdad and AL-Nahrain University, and from 2003 to 2016, he was an assistant professor at Middle Technical University University, Baghdad, Iraq. Presently, he is doing his PhD in the Electrical and Computer Engineering Department at Prairie View A&M University, Prairie View, Texas. He is the author of 25 papers and four books. His main research interests are in the area of sustainable power and energy systems, microgrids, power electronics, and motor drives, the application of digital PID controllers, digital methods to electrical measurements, digital signal processing, and control systems.

1 Continuous and Discrete Signals

Mathematically, signals are represented as a function of one or more independent variables. At this point, we are focusing the attention on signals that involve a single independent variable. Conventionally, it will generally refer to the independent variable as time. There are two types of signals: continuous-time signals and discrete-time signals. In this chapter will focus on the kinds of signals

This chapter gives you a quick way to become familiar with the MATLAB software by introducing you the basic features, commands, and functions. You will discover that entering and solving complex numbers in MATLAB is as easy as entering and solving real numbers, especially with the help of MATLAB built-in complex functions. Upon completion this chapter, and Appendix A you should know how to start MATLAB, how to get HELP, how to assign variables in MATLAB and to perform the typical complex numbers operations (i.e., complex conjugate, addition, subtraction, multiplication, division, expression simplification) and the conversions of complex numbers in both rectangular and polar forms with and without using MATLAB built-in functions.

1.1 CONTINUOUS SIGNALS

The continuous-time signal is the signals or quantities that can be defined and represented at certain time instants of the sequence. A speech signal as a function of time can be classified as a continuous-time signal while the discrete-time signal is the signals or quantities that can be defined and represented at certain time instants of the sequence. The weekly Dow Jones stock market index is an example of a discrete-time signal.

To distinguish between continuous-time and discrete-time signals, we use the symbol "t" to denote the continuous variable and "n" to indicate the discrete-time variable. And for the continuous-time signals, we will enclose the independent variable in parentheses (•), for discrete-time signals we will insert the independent variable in square brackets [•].

A discrete-time signal [nx] may represent a phenomenon for which the independent variable is inherently discrete. Also, a discrete-time signal [nx] may represent successive samples of an underlying aspect for which the independent variable is continuous. For example, the processing of speech on a digital computer requires the use of a discrete-time sequence representing the values of the continuous-time speech signal at discrete points of time. Mostly, it is important to consider signals as related through a modification of the independent variable. These modifications will usually lead the signal to reflection, scaling, and shift.

1.1.1 GENERATION OF CONTINUOUS SIGNALS IN MATLAB

The following is a MATLAB program to generate continuous-time signals like unit step, sawtooth, triangular, sinusoidal, ramp, and sinc function.

```
%generate unit Step
clc;
clear all;
close all;
t=-20:0.01:20;
```

```
L=length(t);
for i=1:L
if t(i)<0
x1(i)=0;
x2(i)=0;
else
x1(i)=1;
x2(i)=t(i);
end;
end;
figure;
plot(t,x1);
xlabel('t');
ylabel('amplitude');
title('unit step');
grid
```

Figure 1.1 shows the output after running the program.

1.1.2 OPERATIONS ON SIGNALS AND SEQUENCES

To perform various operations on signals such as addition, multiplication, scaling, shifting, and folding using the MATLAB program follow the below.

```
%Sum of two signals
clc;
close all;
clear all;
t=0:0.001:1;
L=length(t);
f1=1;
f2=3;
x1=sin(2*pi*f1*t);
```

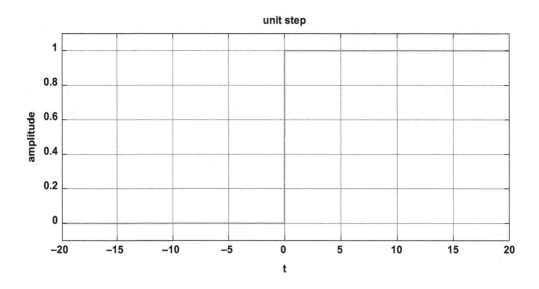

FIGURE 1.1   Generation of the unit step function.

```
x2=sin(2*pi*f2*t);
figure;
subplot(2,1,1);
plot(t,x1,'b',t,x2,'r');
xlabel('t');
ylabel('amplitude');
title('The signals x1(t) and x2(t)');
x3=x1+x2;
subplot(2,1,2);
plot(t,x3);
xlabel('t');
ylabel('amplitude');
title('The sum of x1(t) and x2(t)');
```

Figure 1.2 Shows the results obtained from the sum of the two functions.

```
% Multiplication of two signals
clc;
close all;
clear all;
t=0:0.001:1;
L=length(t);
f1=1;
f2=3;
x1=sin(2*pi*f1*t);
x2=sin(2*pi*f2*t);
figure;
subplot(2,1,1);
plot(t,x1,'b',t,x2,'r');
xlabel('t');
ylabel('amplitude');
title('The signals x1(t) and x2(t)');
x4=x1.*x2;
subplot(2,1,2);
plot(t,x4);
xlabel('t');
ylabel('amplitude');
title('The multiplication of x1(t) and x2(t)');
```

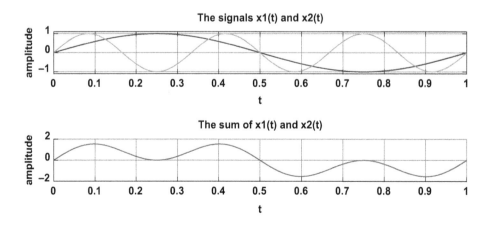

FIGURE 1.2   The sum of functions.

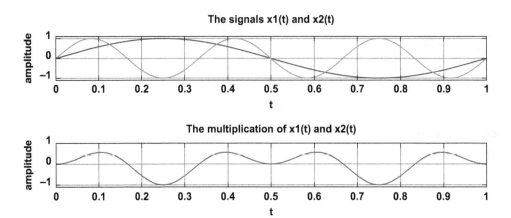

FIGURE 1.3   The multiplying of two functions.

Figure 1.3 shows the results obtained from multiplying the two functions.

```
% Shifting of two signals
clc;
close all;
clear all;
t=0:0.001:1;
L=length(t);
f1=1;
f2=3;
x1=sin(2*pi*f1*t);
x2=sin(2*pi*f2*t);
figure;
subplot(2,1,1);
plot(t,x1,'b',t,x2,'r');
xlabel('t');
ylabel('amplitude');
title('The signals x1(t) and x2(t)');
```

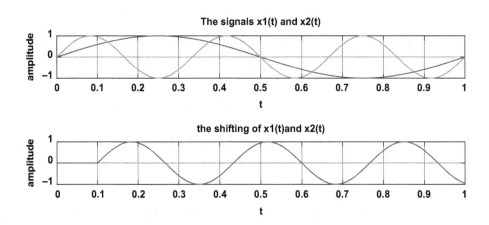

FIGURE 1.4   The Shifting function.

```
x3=[zeros(1,100),x2(1⊙L-100))];
subplot(2,1,2);
plot(t,x3);
title('the shifting of x1(t)and x2(t)');
xlabel('t');
ylabel('amplitude');
```

Figure 1.4 shows the results obtained from shifting the function.

1.2 DISCRETE-TIME SIGNALS

A discrete-time signal is an indexed sequence of real, imaginary, or complex numbers, and it is a function of an integer-valued n that is denoted by $x(n)$. Although the independent variable n need not necessarily represent "time" (n may, for example, correspond to a spatial coordinate or distance), $x(n)$ is generally referred to as a function of time. Figure 1.5 shows a real-valued signal $x(n)$.

In some problems and applications, it is convenient to view $x(n)$ as a vector. Thus, the sequence values $x(0)$ to $x(N-1)$ may often be considered to be the elements of a column vector as follows:

$$X = \left[x(0), x(1),\ldots, x(N-1)\right]^{T} \tag{1.1}$$

Discrete-time signals are often derived from sampling a continuous-time signal, such as speech, with an analog-to-digital (A/D) converter. For example, a continuous-time signal $x(n)$ that is sampled at a rate of $f_s = 1/T_s$ samples per second produces the sampled signal $x(n)$, which is related to $x_a(t)$ as follows:

$$x(n) = x_a\left(nT_s\right) \tag{1.2}$$

1.2.1 COMPLEX SEQUENCES

Generally, a discrete-time signal may be complex-valued. A complex signal can be expressed either in terms of its real or imaginary parts,

$$z(n) = a(n) + j\,b(n) = \operatorname{Re}\{z(n)\} + j\operatorname{Im}\{z(n)\} \tag{1.3}$$

or in the polar form in terms of its magnitude and phase,

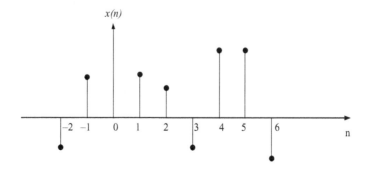

FIGURE 1.5 Generation of a discrete function.

$$z(n) = |z(n)| \exp\left[j \arg\{z(n)\} \right] \qquad (1.4)$$

The magnitude may be derived from the real and imaginary parts as follows:

$$|z(n)|^2 = \text{Re}^2\{z(n)\} + \text{Im}^2\{z(n)\} \qquad (1.5)$$

Whereas the phase may be found using

$$\arg\{z(n)\} = \tan^{-1} \frac{\text{Im}\{z(n)\}}{\text{Re}\{z(n)\}} \qquad (1.6)$$

If $z(n)$ is a complex sequence, the complex conjugate, denoted by $z^*(n)$, is formed by changing the sign on the imaginary part of $z(n)$:

$$z^* = \text{Re}\{z(n)\} - j\,\text{Im}\{z(n)\} = |z(n)| \exp\left[-j \arg\{z(n)\} \right] \qquad (1.7)$$

1.3   SIGNALS AND SYSTEMS

Signals and systems are important for expressing the mathematical model in the modern communication system. Signals are time-varying quantities such as voltages or current. And, the system is a combination of devices and networks (subsystems) that are chosen to perform the desired function. Because of the sophistication of modern communication systems, a great deal of analysis and experimentation with trial subsystems occurs before the actual building of the desired system. Therefore, the communication engineer's tools are the mathematical models for producing signals and systems.

1.4   CLASSIFICATION OF SIGNALS AND SYSTEMS

The signals and systems are classified into:

1.4.1   CONTINUOUS-TIME AND DISCRETE-TIME SIGNALS

By the term *continuous-time* signal, we mean a real or complex function of time $s(t)$, where the independent variable t is continuous.

If t is a discrete variable, i.e., $s(t)$ is defined at discrete times, then the signal $s(t)$ is a *discrete-time* signal. A discrete-time signal is often identified as a sequence of numbers, denoted by $\{s(n)\}$, where n is an integer.

1.4.2   ANALOG AND DIGITAL SIGNALS

If a continuous-time signal $s(t)$ can take on any values in a continuous-time interval, then $s(t)$ is called an *analog signal*.

If a discrete-time signal can take on only a finite number of distinct values, $\{s(n)\}$, then the signal is called a *digital signal*.

1.4.3   DETERMINISTIC AND RANDOM SIGNALS

Deterministic signals are those signals whose values are completely specified for any given time.

Random signals are those signals that take random values at any given times.

1.4.4 Periodic and Nonperiodic Signals

A signal $s(t)$ is a *periodic signal* if $s(t) = s(t + nT_0)$, where T_0 is called the *period* and the *integer* $n > 0$.

If $s(t) \neq s(t + T_0)$ for all t and any T_0, then $s(t)$ is a *nonperiodic* or a *periodic* signal.

1.4.5 Power and Energy Signals

A complex signal $s(t)$ is a power signal if the average normalized power P is finite, where

$$P = \lim_{T \to \infty} \frac{1}{T} \int_{-T/2}^{T/2} s(t)s^*(t)dt \quad 0 < P < \infty \tag{1.8}$$

and $s^*(t)$ is the complex conjugate of $s(t)$.

A complicated signal $s(t)$ is an energy signal if the normalized energy E is finite, where

$$E = \int_{-\infty}^{\infty} s(t)s^*(t)dt = \int_{-\infty}^{\infty} |s(t)|^2 dt \quad 0 < E < \infty \tag{1.9}$$

In communication systems, the received waveform is usually categorized into the desired part that contains the information signal and the undesired part, which is called noise.

1.4.5.1 What Is Digital Signal Processing?

A *signal* is a function of a set of independent variables with time being perhaps the most prevalent single variable. The signal itself carries some kind of information available for observation. By the term *processing*, we mean operating in some fashion on a signal to extract some useful information. In many cases, this processing will be a nondestructive "transformation" of the given data signal; however, some essential processing methods turn out to be irreversible and thus destructive. Also, the word *digital* means that the processing is done with a digital computer or particular purpose digital hardware.

1.4.5.2 Why DSP?

- Rapid advancement in integrated circuit design and manufacture, which leads to more production of powerful DSP system on a single chip with decreasing size and cost.
- Digital processing is inherently stable and reliable.
- In many cases, DSP is used to process several signals simultaneously; it is done by using a technique known as "TDM" (time-division-multiplexing).
- Digital implementation permits easy adjustment of process characteristics during processing, such as that needed for implementing adaptive circuits.

1.4.5.3 Applications (DSP)

- Spectral Analysis
- Speech Recognition
- Biomedical Signal Analysis
- Digital Filtering
- Digital Modems
- Data Encryption
- Image Enhancement and Compression

1.5 INTRODUCTION TO MATLAB IN DSP

MATLAB, which stands for Matrix Laboratory, is a compelling program for performing numerical and symbolic calculations. It is widely used in science and engineering, as well as in mathematics. The basics of the technical language of MATLAB is a technical language to ease scientific computations, and the name derived from Matrix Laboratory. It provides many of the attributes of spreadsheets and programming languages. MATLAB is a case sensitive language (a variable named "c" is different than another one called "C"). In interactive mode, MATLAB scripts are platform independent (right for cross-platform portability). MATLAB works with matrices. Everything MATLAB understands is a matrix (from text to large cell arrays and structure arrays). The MATLAB environment shown in Figure 1.6.

1.5.1 MATLAB Windows

MATLAB works through three basic windows.

Command window: This is the main window. It is characterized by the MATLAB command prompt ">>." When you launch the MATLAB application, the program starts with window and all commands, including those for user-written programs, are typed into this window at the MATLAB prompt.

Graphics window: The output of all graphics commands typed in the command window are flushed to the graphics or figure window, which is a separate gray window with a white background color. The user can create as many windows as the system memory will allow.

Edit window: This is where you write, edit, create, and save your programs in files called M files.

Input–output: MATLAB supports interactive computation taking the input from the screen and flushing the output to the screen. Also, it can read input files and write output files.

Data type: The fundamental data type in MATLAB is the array. It encompasses several distinct data objects-integers, real numbers, matrices, character strings, structures, and cells. There is no need to declare variables as real or complex; MATLAB automatically sets the variable as real.

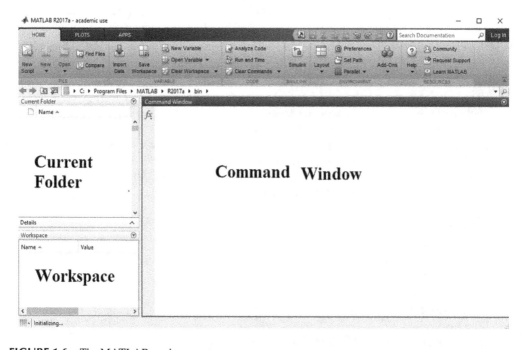

FIGURE 1.6 The MATLAB environment.

Dimensioning: Dimensioning is automatic in MATLAB. No dimension statements are required for vectors or arrays. We can find the dimensions of an existing matrix or a vector with the size and length commands.

All programs and commands can be entered either in the command window or in the M file using the MATLAB editor; then you can save all M files in the folder "work" in the current directory.

1.5.2 BASIC COMMANDS IN MATLAB

No.	Commands	Description
1	$Y = 0:2:20$	This instruction indicates a vector Y which as initial value 0 and final value 20 with an increment of 2. Therefore: $Y = [0\ 2\ 4\ 6\ 8\ 10\ 12\ 14\ 16\ 18\ 20]$.
2	$M = 40:5:100$	$M = [40\ 45\ 50\ 55\ 60\ 65\ 70\ 75\ 80\ 85\ 90\ 95\ 100]$.
3	$N = 0: 1/pi: 1$	$N = [0, 0.3183, 0.6366, 0.9549]$.
4	zeros (1,5)	Creates a vector of one row and five columns whose values are zero Output = [0 0 0 0 0].
5	ones (2,6)	Creates a vector of two rows and six columns Output = 1 1 1 1 1 1 1 1 1 1 1 1.
6	$a = [2\ 2\ -5]$ $b = [3\ 5\ 4]$	$a*b = [6\ 10\ -20]$.
7	plot (t,x) If $x = [6\ 7\ 8\ 9]$ $t = [1\ 2\ 3\ 4]$	This instruction will display a figure window which indicates the plot of x versus t. 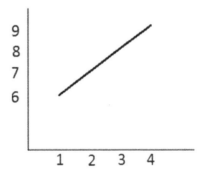
8	stem (t,x)	This instruction will display a figure window as shown.

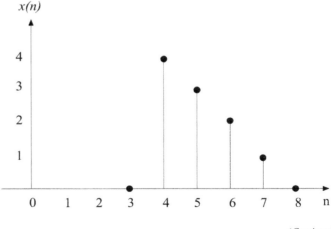

(Continued)

No.	Commands	Description
9	Subplot	This function divides the figure window into rows and columns.
		Subplot (2, 2, 1) divides the figure window into three rows and three columns
		1,2,3,4,…. represent the number of the figure.

(3,3,1)	**(3,3,2)**	**(3,3,3)**
(3,3,4)	**(3,3,5)**	**(3,3,6)**
(3,3,7)	**(3,3,8)**	**(3,3,9)**

No.	Commands	Description
10	Conv	Syntax: $y = \mathrm{conv}(a,b)$
		Description: $y = \mathrm{conv}(a,b)$ convolves vectors a and b.
11	Disp	Syntax: disp(X)
		Description: disp(X) displays an array, without printing the array name. If X contains a text string, the string is displayed.
12	FFT	FFT(X) is the discrete Fast Fourier Transform (FFT) of vector X. For matrices, the FFT operation applied to each column. For N-D arrays, the FFT operation operates on the first non-singleton dimension. FFT(X,N) is the N-point FFT, padded with zeros if X has less than N-points and truncated if it has more.
13	ABS	Absolute value.
		ABS(X) is the absolute value of the elements of X. When X is complex, ABS(X) is the complex modulus (magnitude) of the elements of X.
14	ANGLE	Phase angle.
		ANGLE(H): the phase angles, in radians, of a matrix with complex elements.
15	INTERP	$Y = $ INTERP(X,L) re-samples the sequence in vector X at L times the original sample rate. The resulting resampled vector Y is L times longer, LENGTH(Y) = L*LENGTH(X). Resample data at a higher rate using low-pass interpolation.
16	DECIMATE	$Y = $ DECIMATE(X,M) re-samples the sequence in vector X at 1/M times the original sample rate. The resulting resampled vector Y is M times shorter, LENGTH(Y) = CEIL(LENGTH(X)/M). By default, DECIMATE filters the data with an 8th order Chebyshev Type I low-pass filter with cut-off frequency, 8*(Fs/2) /R, before re-sampling. Resample data at a lower rate after low-pass filtering.
17	xlabel	Syntax: xlabel('string')
		Description: xlabel('string') labels the x-axis of the current axes.
18	ylabel	Syntax: ylabel('string')
		Description: ylabel('string') labels the y-axis of the current axes.
19	Title	Syntax: title('string')
		Description: title('string') outputs the string at the top and in the center of the current axes.
20	grid on	Syntax: grid on.
		Description: grid on adds major grid lines to the current axes.
21	*Help*	List topics on which support is available.
22	*Help command name*	Provides help on the topic selected.
23	*Demo*	Runs the demo program.

(Continued)

No.	Commands	Description
24	*Who*	Lists variables currently in the workspace.
25	*Whos*	Lists variables currently in the workspace with their size.
26	*Clear*	Clears the workspace, all the variables are removed.
32	*Clear x,y,z*	Clears only variables x,y,z.
33	*Qult*	Quits MATLAB.
34	*fir, delay, cas, sos, cas2can*	FIR and IIR filtering.
35	*cfir2, cdelay2, wrap2*	Circular FIR filtering.
36	*dtft*	DTFT computation.
37	*sigav, sg, sgfilt, ecg*	Signal averaging, SG smoothing.
38	*kwind, I0, kparm, kparm2*	Kaiser window.
39	*klh, dlh, kbp, dbp, kdiff, ddiff, khilb, dhilb*	FIR filter design.
40	*lhbutt, bpsbutt, lhcheb1, lhcheb2, bpcheb2, bscheb2*	IIR filter design.

1.6 SOME FUNDAMENTAL SEQUENCES

There are three simple discrete-time signals that are frequently used in the representation and description of more complicated signals. These are the *unit sample*, the *unit step*, and the *exponential*. The unit sample,

Impulse response: Denoted by $\delta(n)$, and defined by

$$\delta(n) = \begin{cases} 1 & n = 0 \\ 0 & \text{otherwise} \end{cases} \tag{1.10}$$

1.6.1 IMPULSE RESPONSE IN MATLAB

To find the impulse response of the given LTI system given in a difference equation form or transfer function form of any order and plot the same. The impulse response of the system generates using the MATLAB program as follow:

```
clc;
clear all;
close all;
num=input('type the numerator vector');
den=input('type the denominator vector');
N= input(' enter the desired length of the output sequence');
n=0:N-1;
imp=[1 zeros(1,N-1)];
H=filter(num, den, imp);
disp(' the impulse response of the system is ');
disp(H);
stem(n,H);
xlabel('n');
ylabel('h(n)');
title(' Impulse response');
```

Ans:
type the numerator vector [1 3 –3 2 5 –2 2 4 –4]

type the denominator vector 1
enter the desired length of the output sequence 12
the impulse response of the system is
1   3   –3   2   5   –2   2   4   –4   0   0   0

The unit step, denoted by $u(n)$, is defined by

$$u(n) = \begin{cases} 1 & n \geq 0 \\ 0 & \text{otherwise} \end{cases} \tag{1.11}$$

and is related to the unit sample by

$$u(n) = \sum_{k=-\infty}^{n} \delta(n) \tag{1.12}$$

Similarly, a unit sample is written as a difference of two steps:

$$\delta(n) = u(n) - u(n-1) \tag{1.13}$$

Finally, an exponential sequence is defined by

$$x(n) = a^n \tag{1.14}$$

where a is a real or complex number. Of particular interest is the exponential sequence that formed when $a = e^{-j\omega_0}$ where ω_0 is a real number. In this case, $x(n)$ is a complex exponential:

$$e^{jn\omega_0} = \cos(n\omega_0) + j\sin(n\omega_0) \tag{1.15}$$

1.6.2   SIGNAL DURATION

Discrete-time signals are conveniently classified in terms of their duration or extent. For example, a discrete-time sequence is said to be a *finite-length sequence* if it is equal to zero for all values of n outside a finite interval $[N_1, N_2]$. Signals that are not finite in length, such as the unit step and the complex exponential, are said to be *infinite-length sequences*. Infinite-length sequences may be further classified as either being right-sided, left-sided, or two-sided. A *right-sided sequence* is any infinite-length sequence that is equal to zero for all values of $n < n_0$ for integer n_0. The unit step is an example of a right-sided sequence. Similarly, an infinite-length sequence $x(n)$ is said to be *left-sided* if, for integer n_0, $x(n) = 0$ for all $n > n_0$. An example of a left-sided sequence is

$$x(n) = u(n_0 - n) = \begin{cases} 1 & n \leq n_0 \\ 0 & n > n_0 \end{cases} \tag{1.16}$$

which is a time-reversed and delayed unit step. An infinite-length signal that is neither right-sided nor left-sided, such as the complex exponential, is referred to as a *two-sided sequence*.

1.7   GENERATION OF DISCRETE SIGNALS IN MATLAB

To write a MATLAB program to generate discrete-time signals like a unit-impulse, unit step, unit ramp, exponential signal, and sinusoidal sequences follow the below.

% Generation of unit impulse, unit step, and unit ramp Sequences

```
clc;
clear all;
close all;
n=-10:1:10;
L=length(n);
for i=1:L
    if n(i)==0
      x1(i)=1;
    else x1(i)=0;
    end;
    if n(i)>=0
      x2(i)=1;
      x3(i)=n(i);
    else x2(i)=0;
      x3(i)=0;
    end;
end;
figure;
stem(n,x1);
xlabel(' n ---->');
ylabel('amplitude---->');
title('Unit-impulse signal');
```

Figure 1.7 shows the generation of a unit-impulse discrete function.

Unit step

```
clc;
  clear all;
  close all;
  n=-5:1:15;
  L=length(n);
  for i=1:L
    if n(i)>=0
      x1(i)=1;
      x2(i)=n(i);
    else
      x1(i)=0;
      x2(i)=0;
    end;
```

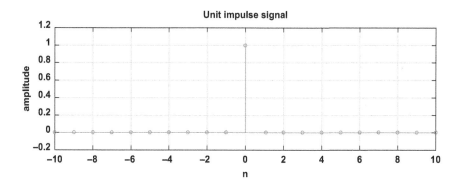

FIGURE 1.7   Generation of unit-impulse discrete function.

```
end;
figure;
stem(n,x1);
xlabel('n');
ylabel('amplitude');
title('Unit step signal')
grid
```

Figure 1.8 shows the generation of a unit-impulse discrete function.

```
clc;
clear all;
close all;
n=-5:1:20;
L=length(n);
for i=1:L
  if n(i)>=0
    x1(i)=1;
    x2(i)=n(i);
  else
    x1(i)=0;
    x2(i)=0;
  end;
end;
stem(n,x2);
xlabel('n');
ylabel('amplitude');
title('Unit ramp signal');
```

Figure 1.9 shows the generation of a unit ramp discrete function.

Generate exponential sequence $a < 1$

```
clc;
clear all;
close all;
n=-10:1:10;
L=length(n);
a=0.7;
x1=a.^n;
```

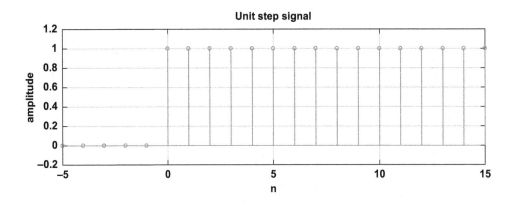

FIGURE 1.8 Generation of unit step discrete function.

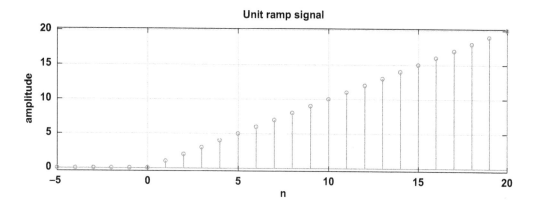

FIGURE 1.9   Generation of unit ramp discrete function.

```
stem(n,x1);
xlabel('n');
ylabel('amplitude');
title('exponential signal');
```

Figure 1.10 shows the generation of an exponential sequence $a < 1$ discrete function.

% Exponential sequence $a > 1$

```
clc;
clear all;
close all;
n=-5:1:15;
L=length(n);
a=1.5;
x1=a.^n;
stem(n,x1);
xlabel('n');
ylabel('amplitude');
title('exponential signal');
```

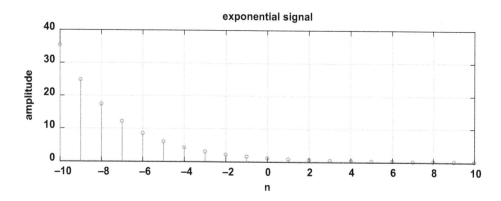

FIGURE 1.10   Shows the generation of exponential sequence $a < 1$ discrete function.

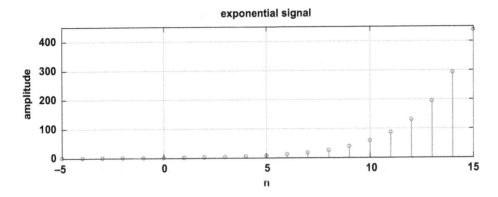

FIGURE 1.11   Generation of exponential sequence $a > 1$ discrete function.

Figure 1.11 shows the generation of an exponential sequence $a > 1$ discrete function.

% Generate exponential sequence $a = 1$

```
clc;
clc;
clear all;
close all;
n=-15:1:15;
L=length(n);
a=1;
x1=a.^n;
stem(n,x1);
xlabel('n');
ylabel('amplitude');
title('exponential signal');
```

Figure 1.12 shows the generation of an exponential sequence $a = 1$ discrete function.

% Generate sinusoidal sequence

```
clc;
clear all;
close all;
```

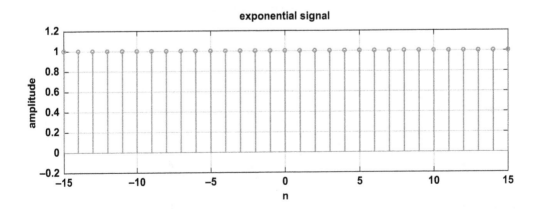

FIGURE 1.12   Generation of exponential sequence $a = 1$ discrete function.

```
n=-20:1:20;
L=length(n);
f=0.1;
x1=sin(2*pi*f*n);
stem(n,x1);
xlabel('n');
ylabel('amplitude');
title('sinusoidal signal');
```

Figure 1.13 shows the generation of a sine discrete function.

PROBLEMS

1.1 Write a MATLAB program to generate a delayed unit step signal with a delay of 8 seconds, then run it and display the result.

1.2 Write a MATLAB program to determine the average normalized power of the triangular signal generated then run the program to display the result.

1.3 Write a program to display the signals:

i $x(t) = t\,u(t) - 3\,t\,u(t-1) + 2\,t\,u(t-1.5)$

ii $f(t) = u\!\left(\sin\!\left(\dfrac{\pi t}{T}\right)\right) - u\!\left(-\sin\!\left(\dfrac{\pi t}{T}\right)\right)$

1.4 Write a MATLAB program to generate:
   - A delayed unit sample sequence with a delay of 13 samples.
   - A delayed unit step sequence with an advance of 7 samples.
   - An exponential sequence with $a = 1.4$.
   - A sinusoidal sequence of frequencies 0.7, 1.3, and 5.
   Run the modified program and display the sequences generated.

1.5 Write MATLAB programs to generate the square wave, and the sawtooth wave sequences, of the types shown in Figure 1.14 a and b. Using these programs, generate and plot the sequences.

1.6 Write MATLAB programs to generate the wave impulse sequences given below. Using these programs, create and plot the sequences.

$$x(n) = u(4-n) = \begin{cases} 4 & n \le 4 \\ 0 & n > 4 \end{cases}$$

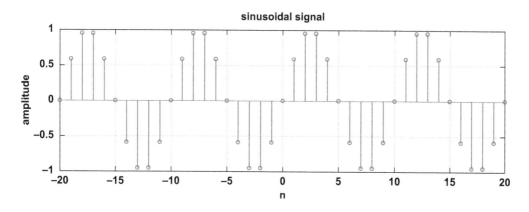

FIGURE 1.13   Generation of sine discrete function.

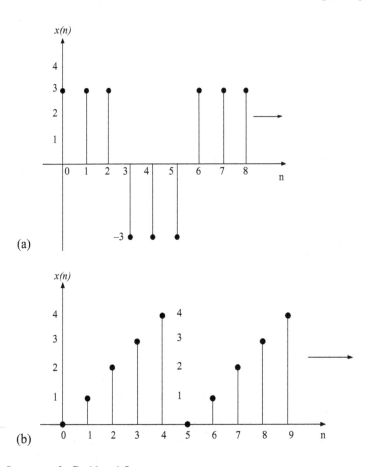

FIGURE 1.14   Sequences for Problem 1.5.

1.7   Write MATLAB programs to generate the wave impulse sequences given below. Using these programs, create and plot the sequences.

$$x(n) = u(2+n) = \begin{cases} 8 & n \le -2 \\ 0 & n > -2 \end{cases}$$

1.8   Write MATLAB programs to generate the wave unit step sequences given below. Using these programs, create and plot the sequences.

$$u(n) = \begin{cases} 1 & n \ge 1 \\ 0 & \text{otherwise} \end{cases}$$

1.9   Write MATLAB programs to generate the wave unit step sequences given below. Using these programs, generate and plot the sequences.

$$u(n) = \begin{cases} 0.75 & n \ge -2 \\ 0 & \text{otherwise} \end{cases}$$

2 Signals Properties

A discrete-time system has inputs and outputs that are discrete-time functions, and a continuous-time system has inputs and outputs that are continuous-time functions. This chapter deals with the signal's properties of discrete signals as Periodic and Aperiodic Sequences, Symmetric Sequences, Even and Odd Parts of a Signal in MATLAB, Signal Manipulations, Transformations of the Independent Variable, Signal Decomposition. Also, the chapter, including System Properties and Linear time-invariant causal systems.

2.1 PERIODIC AND APERIODIC SEQUENCES

A discrete-time signal is organized into two categories. It has to be either *periodic* or *aperiodic*. Suppose a signal $x(n)$ is classified to be periodic, for positive real integer N, and

$$x(n) = x(n+N) \tag{2.1}$$

For all samples n. It resembles the equivalence of repeating itself in the sequence for each of the N samples. If a signal is periodic with period N, it is also periodic with period $2N$, period $3N$, and all other integer multiples of N. If the signal $x(n)$ is not satisfied for any integer N, then $x(n)$ is assumed to be an aperiodic signal.

Example 2.1

The signals

$$x_1(n) = a^n u(n) = \begin{cases} a^n & n \le 0 \\ 0 & n > 0 \end{cases}$$

and $x_2 = \cos(n^2)$ are not periodic, whereas the signal $x_3(n) = e^{j\pi n/8}$ is periodic and has a fundamental period of $N = 16$.

If $x_1(n)$ is a sequence that is periodic with a period of N_1, and $x_2(n)$ is an additional sequence that is periodic with a period of N_2, the sum

$$x(n) = x_1(n) + x_2(n) \tag{2.2}$$

will continuously be periodic, and the fundamental period will be

$$N = \frac{N_1 N_2}{\gcd(N_1, N_2)} \tag{2.3}$$

where $\gcd(N_1, N_2)$ means *the greatest common divisor* of N_1 and N_2. Likewise, the same phenomena are valid for the product; that is

$$x(n) = x_1(n) x_2(n) \tag{2.4}$$

will be periodic with a period of N. However, the first period could be smaller. Given any sequence of $x(n)$, a periodic signal may always be molded by replicating $x(n)$ as follows:

$$y(n) = \sum_{k=-\infty}^{\infty} x(n - kN) \tag{2.5}$$

where N is a positive integer, and in this case $y(n)$ will be periodic with period N.

2.2   EVEN AND ODD PARTS OF A SIGNAL (*SYMMETRIC SEQUENCES*)

A discrete-time signal will often possess some form of symmetry that is manipulated in solving problems. Two symmetries of interest are as follows:

Definition: A real-valued signal is said to be *even* if, for all n,

$$x(n) = x(-n) \tag{2.6}$$

whereas a signal is said to be *odd* if, for all n,

$$x(n) = -x(-n) \tag{2.7}$$

Any signal $x(n)$ is decomposed into a sum of its even part, $x(n)$, and it is an odd part, $x(n)$ as follows:

$$x(n) = x_e(n) + x_o(n) \tag{2.8}$$

To find the even part of $x(n)$, we form the sum

$$x_e(n) = \frac{1}{2}\left\{x(n) + x(-n)\right\} \tag{2.9}$$

whereas to find the odd part we take the difference

$$x_o(n) = \frac{1}{2}\left\{x(n) - x(-n)\right\} \tag{2.10}$$

For complex sequences, the symmetries of interest are slightly different.

A complex signal is said to be *conjugate symmetric (hermitian)* if, for all n,

$$x(n) = x^*(-n) \tag{2.11}$$

and a signal is said to be *conjugate antisymmetric* if, for all n,

$$x(n) = -x^*(-n) \tag{2.12}$$

Any complex signal may always be decomposed into a sum of a conjugate symmetric signal and a conjugate antisymmetric signal.

Follow the MATLAB program below to find the even and odd parts of a signal.

```
clc;
clear all;
close all;
t=-10:0.1:10;
A=0.8;
x1=sin(t);
x2=sin(-t);
if(x2==x1)
disp('The given signal is even signal');
else
if(x2==(-x1))
disp('The given signal is odd signal');
else
disp('The given signal is neither even nor odd');
end
end
xe=(x1+x2)/2;
xo=(x1-x2)/2;
subplot(2,2,1);
```

```
plot(t,x1);
xlabel('t');ylabel('x(t)');title('signal x(t)');
subplot(2,2,2);
plot(t,x2);
xlabel('t');ylabel('x(t)');title('signal x(-t)');
subplot(2,2,3);
plot(t,xe);
xlabel('t');ylabel('x(t)');title('even part signal x(t)');
subplot(2,2,4);
plot(t,xo);
xlabel('t');ylabel('x(t)');title('odd part signal x(t)');
```

Figure 2.1 shows an odd signal after running the program.

```
clc;
clear all;
close all;
t=-10:0.1:10;
A=0.8;
x1=cos(t);
x2=cos(-t);
if(x2==x1)
disp('The given signal is even signal');
else
if(x2==(-x1))
disp('The given signal is odd signal');
else
disp('The given signal is neither even nor odd');
end
end
xe=(x1+x2)/2;
xo=(x1-x2)/2;
subplot(2,2,1);
plot(t,x1);
xlabel('t');ylabel('x(t)');title('signal x(t)');
subplot(2,2,2);
plot(t,x2);
xlabel('t');ylabel('x(t)');title('signal x(-t)');
subplot(2,2,3);
```

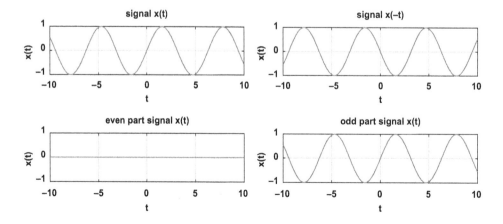

FIGURE 2.1   Odd signal.

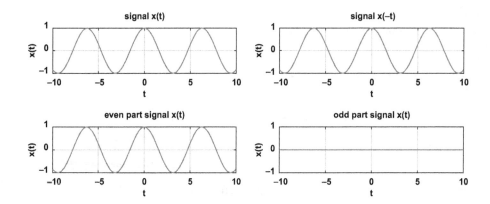

FIGURE 2.2   Even signal.

```
plot(t,xe);
xlabel('t');ylabel('x(t)');title('even part signal x(t)');
subplot(2,2,4);
plot(t,xo);
xlabel('t');ylabel('x(t)');title('odd part signal x(t)');
```

Figure 2.2 shows an even signal after running the program.

```
clc;
clear all;
close all;
t=-10:0.1:10;
K=1.2;
x1=K.^(t);
x2=K.^(-t);
if(x2==x1)
disp('The given signal is even signal');
else
if(x2==(-x1))
disp('The given signal is odd signal');
else
disp('The given signal is neither even nor odd');
end
end
xe=(x1+x2)/2;
xo=(x1-x2)/2;
subplot(2,2,1);
plot(t,x1);
xlabel('t');ylabel('x(t)');title('signal x(t)');
subplot(2,2,2);
plot(t,x2);
xlabel('t');ylabel('x(t)');title('signal x(-t)');
subplot(2,2,3);
plot(t,xe);
xlabel('t');ylabel('x(t)');title('even part signal x(t)');
subplot(2,2,4);
plot(t,xo);
xlabel('t');ylabel('x(t)');title('odd part signal x(t)');
```

Figure 2.3 shows neither an even nor an odd signal after running the program.

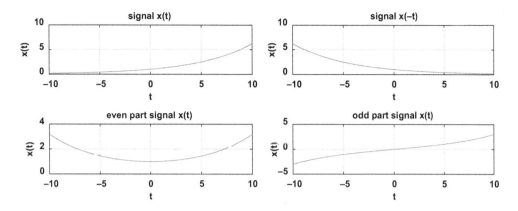

FIGURE 2.3 Neither even nor odd signal.

2.3 SIGNAL MANIPULATIONS

The study of discrete-time signals and systems, will be concerned with the manipulation of signals. These manipulations are commonly compositions of a few necessary signal transformations.

2.3.1 Transformations of the Independent Variable

Sequences are often altered and manipulated by modifying the index n as follows:

$$y(n) = x(f(n)) \tag{2.13}$$

Where $f(n)$ is a function of n, there are some values of n which gives its role an undefined state, specifically if $f(n)$ is not an integer then $y(n) = x(f(n))$ is perceived as undefined. The effects of modifying the index n can be achieved by using a simple tabular approach of listing, where each value for n will give the setting $y(n) = x(f(n))$. The most common transformations of the independent variables are noted as shifting, reversal, and scaling, which are defined below.

2.3.1.1 Shifting

Shifting is the transformation defined by the notation $f(n) = n - n_o$. For the statement, $y(n) = x$ $(n - n_o)$, if n_0 is positive (referred to as a delay), then $x(n)$ is shifted to the right by n_o samples. Likewise, if n_0 is negative (referred to as an advance), then it will be shifted to the left by n_o samples (referred to as an advance).

2.3.1.2 Reversal

This transformation is estimated by the statement $f(n) = -n$ which involves "flipping" the signal $x(n)$ concerning the index n.

2.3.1.3 Time-Scaling

This transformation is defined by the either the notation $x(n) = Mn$ or $f(n) = n/N$ where M and N are positive integers. In the case of $f(n) = Mn$, the sequence $x(Mn)$ is formed by taking every M^{th} sample of $x(n)$ (this operation is known as down-sampling). For the statement, $f(n) = n/N$ the sequence $y(n) = x(f(n))$ is defined as follows:

$$y(t) = \begin{cases} x\left(\dfrac{n}{N}\right) & n = 0, \pm N, \pm 2N, \cdots \\ 0 & \text{otherwise} \end{cases} \tag{2.14}$$

(This operation is known as up-sampling.)

Shifting, reversal, and time-scaling operations are perceived as order-dependent. For example, see the two systems that are shown in Figure 2.4. One of them represents a delay followed by a reversal, and then the other one shows a reversal followed by a delay. As identified before, the outputs of these two systems are not the same.

2.3.1.4   Addition, Multiplication, and Scaling

The most common types of amplitude transformations are *addition*, *multiplication*, and *scaling*.

2.3.1.5   Addition

The sum of two signals

$$y(n) = x_1(n) + x_2(n) \quad -\infty < n < \infty \tag{2.15}$$

is formed by the pointwise addition of the signal values.

2.3.1.6   Multiplication

The multiplication of two signals

$$y(n) = x_1(n)x_2(n) \quad -\infty < n < \infty \tag{2.16}$$

is formed by the pointwise product of the signal values.

2.3.1.7   Scaling

Amplitude scaling of a signal $x(n)$ by a constant c is accomplished by multiplying every signal value by c:

$$y(n) = c\,x(n) \quad -\infty < n < \infty \tag{2.17}$$

This operation may also be considered to be the product of two signals, $x(n)$ and $f(n) = c$.

2.3.1.8   Signal Decomposition

The unit sample may be used to decompose an arbitrary signal $x(n)$ into a sum of weighted and shifted unit samples as follows:

$$x(n) = \cdots + x(-1)\delta(n+1) + x(0)x(-1)\delta(n) + x(1)\delta(n-1) + x(2)\delta(n-2) + \cdots \tag{2.18}$$

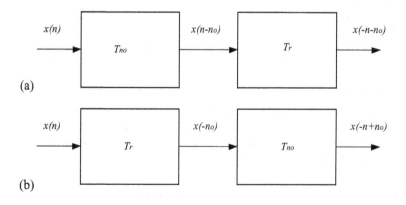

FIGURE 2.4   Time-scaling operation. (a) A delay T_{no} followed by a time-reversal Tr. (b) A time-reversal Tr followed by a delay T_{no}.

This decomposition may be written concisely as:

$$x(n) = \sum_{k=-\infty}^{\infty} x(k)\delta(n-k) \qquad (2.19)$$

where each term in the sum, $x(n)\,\delta(n-k)$, is a signal that has an amplitude of $x(k)$ at the time $n = k$ and a value of zero for all other values of n. This decomposition is the discrete version of the **sifting property** for continuous-time signals and used in the derivation of the convolution sum.

2.4   DISCRETE-TIME SYSTEMS

A discrete-time system is a mathematical operator or mapping that transforms one signal (the input) into another signal (the output) using a fixed set of rules or operations. The notation $T[.]$ is used to represent a general system as shown in Figure 2.5, in which an input signal $x(n)$ is transformed into an output signal $y(n)$ through the transformation $T[.]$:

$$y(n) = x^2(n)$$

or

$$y(n) = 0.5\,y(n-1) + x(n)$$

However, it could be described as a system in terms of an algorithm that provides a sequence of instructions or operations that is to apply to the input signal, such as:

$$y_1(n) = 0.5\,y_1(n-1) + 0.25\,x(n)$$

$$y_2(n) = 0.25\,y_2(n-1) + 0.5\,x(n)$$

$$y_3(n) = 0.4\,y_3(n-1) + 0.5\,x(n)$$

$$y(n) = y_1(n) + y_2(n) + y_3(n)$$

2.4.1   SYSTEM PROPERTIES

2.4.1.1   Memoryless System

The first property is concerned with whether or not a system has memory. A system is said to be memoryless if the output at any time of $n = n_o$ depends only on the input at a time of $n = n_o$. In other words, a system is memoryless if, for any n_o, we can determine the value of $y(n_o)$ given only the value of $x(n_o)$.

For example, the system $y(n) = x^2(n)$ is memoryless because $y(n_o)$ depends only on the value of $x(n_o)$ at the time n_o. The system $y(n) = x(n) + x(n-1)$ on the other hand, is not memoryless because the output at the time n_o depends on the value of the input both at the time and at the time $(n_o - 1)$.

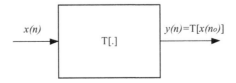

FIGURE 2.5   Discrete-time system.

2.4.1.2 Additivity

An additive system is one for which the response to a sum of inputs is equal to the sum of the inputs individually. Thus, A system is said to be additive if $T[x_1(n) + x_2(n)] = T[x_1(n)] + T[x_2(n)]$ for any signals $x_1(n)$ and $x_2(n)$.

2.4.1.3 Homogeneity

A system is said to be homogeneous if scaling the input by the constant results in a scaling of the output by the same amount. Specifically, a system is said to be homogeneous if $T[cx(n)] = cT[x(n)]$ for any *complex* constant c and any input sequence $x(n)$.

Example 2.2

The system defined by

$$y(n) = \frac{x^2(n)}{x(n-1)}$$

which is not additive because

$$T[x_1(n) + x_2(n)] = \frac{(x_1(n) + x_2(n))^2}{x_1(n-1) + x_2(n-1)}$$

which is not the same as

$$T[x_1(n) + x_2(n)] = \frac{(x_1(n))^2}{x_1(n-1)} + \frac{(x_2(n))^2}{x_2(n-1)}$$

This system is, however, homogeneous because, for an input $cx(n)$ the output is

$$T[cx(n)] = \frac{(cx(n))^2}{cx(n-1)} = c\frac{(x(n))^2}{x(n-1)} = cT[x(n)]$$

On the other hand, the system defined by the equation $y(n) = x(n) + x^*(n)$ is additive because

$$[x_1(n) + x_2(n)] + [x_1(n-1) + x_2(n-1)]^* = [x_1(n) + x_1^*(n-1)] + [x_2(n) + x_2^*(n-1)]$$

However, this system is not homogeneous because the response to $cx(n)$ is

$$T[cx(n)] = cx(n) + c^*x^*(n-1)$$

which is not the same as

$$cT[x(n)] = cx(n) + cx^*(n-1)$$.

2.4.1.4 Stability

In many applications, it is essential for a system to have a response, $y(n)$, that is bounded in amplitude whenever the input is bounded. A system with this property is said to be *stable* in the bounded input/bounded output (BIBO) sense. Specifically, a system is said to be stable in the defined input/bounded output sense if, for any input that is bounded, $|x(n)| \le A < \infty$, the output will be bounded, $|y(n)| \le B < \infty$.

For a linear shift-invariant system, stability is guaranteed if the unit sample response is summable:

$$\sum_{n=-\infty}^{\infty} |h(n)| < \infty$$

Example 2.3

A system with unit sample response $h(n) = a^n u(n)$ will be stable whenever $|a| < 1$ because

$$\sum_{n=-\infty}^{\infty} |h(n)| = \sum_{n=0}^{\infty} |a|^n = \frac{1}{1-|a|} \quad |a| < 1$$

The system is described by the equation $y(n) = nx(n)$. On the other hand, it is not described as stable because the response to a unit step, $x(n) = u(n)$, is $y(n) = nu(n)$, which is unbounded.

2.5   LINEAR TIME-INVARIANT CAUSAL SYSTEMS (LTI)

Linear time-invariant systems are the most important systems in DSP fields. Besides the previously mentioned properties, the most important features of this system are linearity, time-invariance, and causality.

2.5.1   LINEARITY

A system that is both additive and homogeneous is said to be *linear*. Thus, a system is said to be linear if $T[a_1 x_1(n) + a_2 x_2(n)] = a_1 T[x_1(n)] + a_2 T[x_2(n)]$ for any two inputs $x_1(n)$ and $x_2(n)$ and any complex constants a_1 and a_2.

Linearity greatly simplifies the evaluation of the response of a system to a given input. For example, given $x(n)$:

$$x(n) = \sum_{k=-\infty}^{\infty} x(k)\,\delta(n-k) \tag{2.20}$$

and using the additivity property it follows that the output $y(n)$ is written as

$$y(n) = T[x(n)] = T\left[\sum_{k=-\infty}^{\infty} x(k)\,\delta(n-k)\right] = \sum_{k=-\infty}^{\infty} T[x(k)\,\delta(n-k)] \tag{2.21}$$

Because the coefficients $x(k)$ are constants, we may use the homogeneity property to write:

$$y(n) = T\left[\sum_{k=-\infty}^{\infty} x(k)\,\delta(n-k)\right] \sum_{k=-\infty}^{\infty} x(k)T[\delta(n-k)] \tag{2.22}$$

If we define $h_k(n)$ as the response of the system to a unit sample at time $n = k$,

$$h_k(n) = T[\delta(n-k)] \tag{2.23}$$

$$y(n) = \sum_{k=-\infty}^{\infty} x(k)\,h_k(n) \tag{2.24}$$

which is known as the *superposition summation*.

For example, consider the digital filters as:

1) $y[n] = x[n] + x[n-1] + x[n-2] + \cdots$ as represented in Figure 2.6.
2) $y[n] = y[n-1] + x[n]$ as represented in Figure 2.7.
3) $y[n] = 1.8\,y[n-1] - 0.9\,y[n-2] + x[n] - 1.9\,x[n-1] + x[n-2]$ as represented in Figure 2.8.

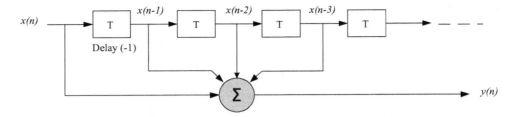

FIGURE 2.6 $y[n] = x[n] + x[n-1] + x[n-2] + \cdots$ representation.

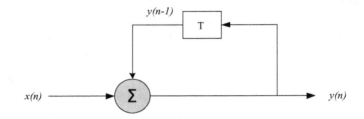

FIGURE 2.7 $y[n] = y[n-1] + x[n]$ representation.

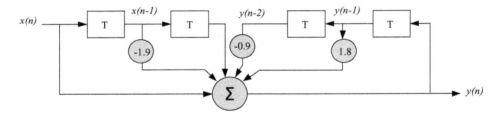

FIGURE 2.8 $y[n] = 1.8\,y[n-1] - 0.9\,y[n-2] + x[n] - 1.9\,x[n-1] + x[n-2]$ representation.

Example 2.4

Find the first four samples values of the impulse response $h[n]$ for each the following digital processors:

a) The system illustrated in Figure 2.9.
b) The system $y[n] = x[n] + x[n-1] + x[n-2] + \cdots$

SOLUTION

a) From Figure 2.9, the output equation is given as

$$y[n] = -0.7\,y[n-1] + x[n]$$

The impulse response is

$$h[n] = -0.7\,h[n-1] + \delta[n]$$

The system is causal, so $h[n] = 0$ for all $n < 0$.
For $n = 0$,

$$h[0] = -0.7\,h[-1] + \delta[0] = 0 + 1 = 1$$

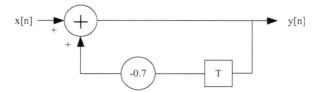

FIGURE 2.9 System for Example 2.4.

$n = 1$,
$$h[1] = -0.7\,h[0] + \delta[1] = -0.7*1+0 = -0.7$$

$n = 2$,
$$h[2] = -0.7h[1] + \delta[2] = (-0.7)*(-0.7)+0 = 0.49$$

$n = 3$,
$$h[3] = -0.7\,h[2] + \delta[3] = (-0.7)*(0.49)+0 = -0.343$$

b) $y[n] = x[n] + x[n-1] + x[n-2] + \cdots$

$h[n] = \delta[n] + \delta[n-1] + \delta[n-2] + \cdots$

$h[0] = \delta[0] + \delta[-1] + \delta[-2] + \cdots = 1$

$h[1] = \delta[1] + \delta[0] + \delta[-1] + \cdots = 1$

$h[2] = \delta[2] + \delta[1] + \delta[0] + \cdots = 1$

$h[3] = \delta[3] + \delta[2] + \delta[1] + \cdots = 1$

$\cdots$

Example 2.5

Find the first five sample values of the impulse response $h[n]$ for each system illustrated in Figure 2.10

1) The impulse response is
$$h[n] = 0.5h[n-1] + \delta[n]$$

The system is causal, so $h[n] = 0$ for all $n < 0$.
For $n = 0$,
$$h[0] = 0.5\,h[-1] + \delta[0] = 0+1 = 1$$

$n = 1$,
$$h[1] = 0.5\,h[0] + \delta[1] = -0.7*1+0 = 0.5$$

$n = 2$,
$$h[2] = 0.5h[1] + \delta[2] = (0.5)*(0.5)+0 = 0.25$$

$n = 3$,
$$h[3] = 0.5\,h[2] + \delta[3] = (0.5)*(0.25)+0 = 0.125$$

$n = 4$,
$$h[4] = 0.5\,h[3] + \delta[2] = (0.5)*(0.125)+0 = 0.0625$$

$\cdots$

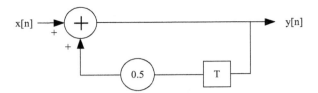

FIGURE 2.10   System for Example 2.5.

2) For the step response

$$s[n] = \sum_{m=-\infty}^{n} h(m)$$

$$s[0] = h[0] = 1$$

$$s[1] = h[0] + h[1] = 1.5$$

$$s[2] = h[0] + h[1] + h[2] = 1.75$$

$$s[3] = h[0] + h[1] + h[2] + h[3] = 1.875$$

$$s[4] = h[0] + h[1] + h[2] + h[3] + h[4] = 1.9375$$

…and so on.

Example 2.6

For the sequence following determine the linearity of each system, where $x(n)$ is the input and $y(n)$ is the output.

a) $y(n) = ln(x(n))$
b) $y(n) = 1 + x(n) + x(n + 1) + x(n + 2)$
c) $y(n) = x(n) + [x(n + 1) x(n - 2)]/x(n)$
d) $y(n) = x(n) \sin(n\pi/2)$

SOLUTION

Use $y(n) = T[cx(n)] = cT[x(n)]$ for a linear system.

a) $y(n) = ln(x(n)) \rightarrow y(n) = ln(cx(n)) = ln\ c + ln\ x(n) \neq cT[x(n)]$, so the system is non-linear.
b) $y(n) = 1 + x(n) + x(n + 1) + x(n + 2)$

$$y_1(n) = 1 + x_1(n) + x_1(n + 1) + x_1(n + 2)$$

$$= 1 + c\ \big[x(n) + x(n + 1) + x(n + 2)\big]$$

but $c\ y(n) = c\ [1 + x(n) + x(n + 1) + x(n + 2)]$
so $y_1(n) \neq c\ y(n)$, so the system is non-linear.

c) $y(n) = x(n) + [x(n + 1) x(n - 2)]/x(n)$
$y_1(n) = x_1(n) + [x_1(n + 1) x_1(n - 2)]/x_1(n)$

$$y_1(n) = cx(n) + \big[cx(n + 1)cx(n - 2)\big] / cx(n)$$

$$= c\big[x(n) + \big[x(n + 1)\ x(n - 2)\big]/x(n)\big] = cy(n)$$

Since $y_1(n) = c\ y(n)$, the system is linear.

  d) $y(n) = x(n) \sin(n\pi/2)$
     Let $x(n) = a_1 x_1(n) + a_2 x_2(n)$

$$y(n) = a_1 x_1(n)\sin(n\pi/2) + a_2 x_2(n)\sin(n\pi/2)$$

$$= a_1 y_1(n) + a_2 y_2(n)$$

so the system is linear.

2.5.2   TIME-INVARIANCE

If a system has the property that a shift (delay) in the input by n_o results in a shift in the output by n_o, the system is said to be time-invariant.

Let $y(n)$ be the response of a system to an arbitrary input $x(n)$. The system is said to be time-invariant if, for any delay of n_o, the response to $x(n - n_o)$ is $y(n - n_o)$. A system that is not time-invariant is said to be time-varying.

In effect, a system will be time-invariant if its properties or characteristics do not change with time. To test for shift-invariance, one needs to compare $y(n - n_o)$ to $T[x(n - n_o)]$. If they are the same for any input $x(n)$ and all shifts n_o, the system is shift-invariant.

Example 2.7

Determine whether each of the following systems is time-invariant.

  a) $y(n) = x^2(n)$

  b) $y(n) = x(n) + x(-n)$

SOLUTION

  a) If $y(n) = x^2(n)$ is the response of the system to $x(n)$, the response of the system to $x'(n) = x(n - n_0)$ is $y'(n) = [x'(n)]^2 = x^2(n - n_0)$.
     Because $y'(n) = y(n - n_0)$, the system is time-invariant.

  b) First, note that the system's response to the input $x(n) = \delta(n)$ is:

$$y(n) = \delta(n) + \delta(-n) = 2\delta(n)$$

whereas the response to $x(n - 1) = \delta(n - 1)$ is

$$y'(n) = \delta(n - 1) + \delta(-n - 1)$$

Because this is not the same as $y(n - 1) = 2\delta(n - 1)$, the system is time-variant.

Example 2.8

For each sequence in the following Determine whether or not the systems are shift-invariant.

  a) $y(n) = x(n) + x(n - 1) + x(n - 2) + x(n - 3)$.

  b) $y(n) = 3x(n)u(n)$.

  c) $y(n) = x(-n)$.

  d) $y(n) = nu(n)$.

  e) $y(n) = \displaystyle\sum_{k=-\infty}^{n} x(k)$

SOLUTION

a) $y(n) = x(n) + x(n-1) + x(n-2) + x(n-3)$

When a response is shifted by n_0:

$$y(n-n_0) = x(n-n_0) + x(n-n_0-1) + x(n-n_0-2) + x(n-n_0-3)$$

The response of the system to $x_1 = x(n-n_0)$ is

$$y_1(n) = x_1(n) + x_1(n-1) + x_1(n-2) + x_1(n-3)$$
$$= x_1(n-n_0) + x_1(n-n_0-1) + x_1(n-n_0-2) + x_1(n-n_0-3)$$

Since $y_1(n) = y(n-n_0)$, the system is shift-invariant.

b) $y(n) = 3x(n)u(n)$

$$y(n) = 3x(n)f(n)$$

where $f(n)$ is the varying function. Systems of this form are always shift-varying provided $f(n)$ is not a constant. So, the sequence $y(n) = 3x(n)u(n)$ is shift-varying.

c) $y(n) = x(-n)$

$$y_1(n) = x_1(-n) = x(n-n_0)$$

$$y(n-n_0) = x\left(-(n-n_0)\right) = x(-n+n_0)$$

which is not equal to $y_1(n)$. Therefore, the system is shift-varying.

d) $y(n) = nu(n)$.

Shift the input $x_1(n-n_0)$

$$y_1(n) = (n-n_0) \cdot u(n-n_0)$$

$$y(n-n_0) = (n-n_0) \cdot u(n-n_0)$$

which is equal to $y_1(n)$. Therefore, the system is shift-invariant.

e) $y(n) = \displaystyle\sum_{k=-\infty}^{n} x(k)$

Shifting the input $x_1(n-n_0)$

$$y_1(n) = \sum_{k=-\infty}^{n} x(k-n_0) = \sum_{k=-\infty}^{n-n_0} x(k)$$

Since this is equal to $y(n-n_0)$, *the* system is shift-invariant.

2.5.3 CAUSALITY

A system property that is important for real-time applications is *causality*, which is defined as a system is said to be *causal* if, for any n_o, the response of the system at a time of n_o depends only on the input up to a time of $n = n_o$.

For a causal system, changes in the output cannot precede changes in the input. Thus, if $x_1(n) = x_2(n)$ for $n \leq n_o$, $y_1(n)$ must be equal to $y_2(n)$ for $n \leq n_o$. A LTI system will be causal if and only if $h(n)$ is equal to zero for $n < 0$.

Example 2.9

The system described by the equation $y(n) = x(n) + x(n - 1)$ is causal because of the value of the output at any time $n = n_o$ depends only on the input $x(n)$ at the time n_o and at the time $n_o - 1$. The system described by $y(n) = x(n) + x(n + 1)$ on the other hand, is noncausal because of the output at time $n = n_o$ depends on the value of the input at the time $n_o - 1$.

Example 2.10

Given the following linear systems,

    a) $y(n) = 0.9 x(n) + 2.2 x(n - 2),$ for $n \geq 0$

    b) $y(n) = 1.25 x(n - 1) + 2.5 x(n + 1) - 4.4 y(n - 1),$ for $n \geq 0$

    determine whether each is causal.

SOLUTION

    a) Since for $n \geq 0$, the output $y(n)$ depends on the current input $x(n)$ and its past value $x(n - 2)$, the system is causal.

    b) Since for $n \geq 0$, the output $y(n)$ depends on the current input $x(n)$ and its future value $x(n + 2)$, the system is noncausal.

Example 2.11

The system is described by the sequence:
    $x(n) = (5 - n)[u\{n\} - u(n - 5)]$. Determine and draw the following:

    a) $y(n) = x(4 - n)$.
    b) $g(n) = x(2n - 3)$.
    c) $h(n) = x(6 - 2n)$.
    d) $v(n) = x(n^2 - 2n + 1)$.

SOLUTION

The system described by the sequence: $X(n) = (5 - n)[u\{n\} - u(n - 5)]$, so the sequence is illustrated as shown in Figure 2.11a.

Example 2.12

For the system described by the sequence shown in Figure 2.12, determine and draw

    a) $x(n - 2)$
    b) $x(-n)$
    c) $x(2n)$
    d) $x(n/2)$

SOLUTION

Figure 2.13 shows the solution of Example 2.12, where Figure 2.13a represents the sequence when there is a delay of $n_0 = 2$. Figure 2.13b represents time-reversal of the sequence while Figure 2.13c illustrates the down-sampling by a factor of 2, and Figure 2.13d represents an up-sampling by a factor of 2.

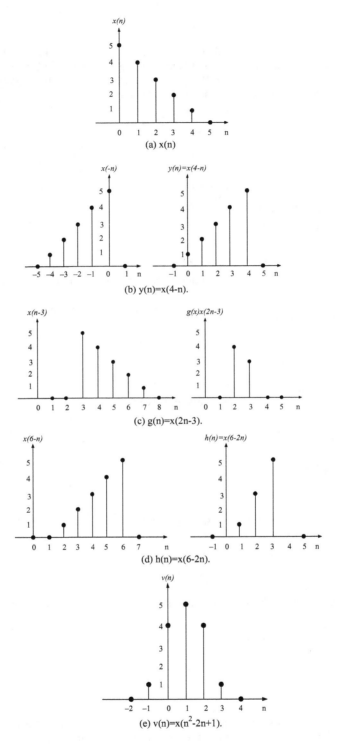

FIGURE 2.11   Sample sequences for Example 2.11.

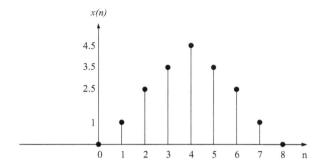

FIGURE 2.12 Sample sequences for Example 2.12.

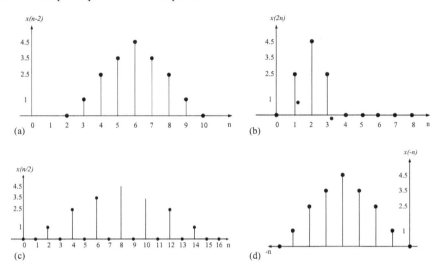

FIGURE 2.13 Sample sequences for the solution of Example 2.12.

Example 2.13

Determine whether the following the sequences are periodic or aperiodic; for each periodic sequence determine the fundamental period.

a) $x(n) = \sin(0.125\pi n)$
b) $x(n) = \cos(\pi + 0.5n)$
c) $x(n) = \sin(n\pi/15)\, e^{j\pi n/16}$
d) $x(n) = Re\{e^{j\pi n/18}\} + Im\{e^{j\pi n/12}\}$

SOLUTION

a) $x(n) = \sin(0.125\pi n)$

$$0.125\pi n = \frac{2\pi n}{N}$$

where N is the sample for one period
$N = 16$ integer number, so the sequence is periodic.
And the fundamental period is 16.

b) $x(n) = \cos(\pi + 0.5n)$

$$0.5n = \frac{2\pi n}{N}$$

$N = 4\pi$ has no integer value, so the sequence is aperiodic.

c) $x(n) = \sin(n\pi/15)\, e^{j\pi n/16}$

$$\frac{n\pi}{15} = \frac{2\pi n}{N_1} \rightarrow N_1 = 30$$

$$\frac{n\pi}{16} = \frac{2\pi n}{N_2} \rightarrow N_2 = 32$$

$$N = \frac{N_1 N_2}{\gcd(N_1, N_2)} = \frac{30 \times 32}{2} = 480$$

N is an integer, so the sequence is periodic and the fundamental period is $N = 480$.

d) $x(n) = Re\{e^{j\pi n/18}\} + Im\{e^{j\pi n/12}\}$

$$\frac{n\pi}{18} = \frac{2\pi n}{N_1} \rightarrow N_1 = 36$$

$$\frac{n\pi}{12} = \frac{2\pi n}{N_2} \rightarrow N_2 = 24$$

$$N = \frac{N_1 N_2}{\gcd(N_1, N_2)} = \frac{36 \times 24}{2} = 432$$

N is an integer, so the sequence is periodic and fundamental period $N = 432$.

2.6   DEFINITIONS

2.6.1   CONTINUOUS-TIME SYSTEM

In continuous systems, we use the system function $H(s)$ or $H(j\omega)$ to describe the frequency response. The system function describes the response of a system to the special input $x(s,t) = e\,st$. The response is $y(s,t) = H(s)e\,st$.

Consider a continuous-time system that is described by a differential equation:

$$y(t) + a_1 y(1)(t) + a_2 y(2)(t) + \cdots + apy(p)(t)$$
$$= b_0 x(t) + b_1 x(1)(t) + \cdots + bqx(q)(t) \tag{2.25}$$

Let us substitute $x(t) = est$ and $y(t) = H(s)est$. Then $x(k)(t) = skest$ and
$y(t) = skH(s)ewt$, and we find

$$\left(1 + a_1 s + a_2 s_2 + \cdots + apsp\right) H(s)e\,st$$
$$= \left(b_0 + b_1 s + \cdots + bqsq\right)e\,st \tag{2.26}$$

This equation will be valid for all values of t provided the coefficients of $e\,st$ on each side of the equation are equal. Solve for the system function:

$$H(s) = \frac{b_0 + b_1 s + \cdots + b_q s^q}{1 + a_1 s + a_2 s^2 + \cdots + a_p s^p} \tag{2.27}$$

This function is valid for all values of s for which the denominator is nonzero, Including complex and imaginary values. It tells us that any system that is described by a linear differential equation with constant coefficients has a system function that is a ratio of polynomials in s. The coefficients are the coefficients from the differential equation.

Consider a continuous-time linear system that is described by an impulse response function $h(t)$. The response of the system to any input $x(t)$ is

$$y(t) = \int_{-\infty}^{\infty} H(\tau)x(t-\tau)d\tau \tag{2.28}$$

Now let $x(t) = e\ st$ and $y(t) = H(s)e\ st$. Then we must have

$$H(s)e^{st} = \int_{-\infty}^{\infty} h(\tau)e^{s(t-\tau)}d\tau = \int_{-\infty}^{\infty} h(\tau)e^{-s\tau}d\tau$$

Therefore,

$$H(t) = \int_{-\infty}^{\infty} h(\tau)e^{-st}d\tau \tag{2.29}$$

This result shows that the system function and the impulse response are a transform pair for time-invariant continuous-time linear systems. Either function is just an alternative way of looking at the behavior of the system.

A significant use for the system function is in describing the behavior of the system in the frequency domain. Every sinusoidal waveform is written in terms of exponentials.

$$\text{Cos}(\omega t) = [e(j\omega t) + e(-j\omega t)]/2 \tag{2.30}$$

Hence, the response to $x(t) = \cos(\omega t)$ is $y(t) = [H(j\omega)ej\omega t + H(-j\omega)e\ -j\omega t]/2$.

For a linear system with real parameters, such as real coefficients in the differential equation or real impulse response, the response to a real input, such as a cosine wave, must also be real. This requires that the system function has conjugate symmetry. $H(-j\omega) = H^*(j\omega)$. Every complex number is written in a polar coordinate form. If Z is a complex number, then it can be written in the form $Z = |Z|\ e\ j\theta$. We can apply this to the system function since it is just a complex number at any fixed value of ω. Hence the output is $y(t) = [|H(j\omega)|ej(\omega t + \theta) + |H(j\omega)|e - j(\omega t + \theta)]/2 = |H(j\omega)|\cos(\omega t + \theta)$.

Therefore, $|H(j\omega)|$ is the "gain" of the system at frequency ω and θ is the phase shift at frequency ω.

2.6.2 Discrete-Time System

The system function for a discrete-time system is based on the difference equation rather than the differential equation. Just as in a continuous system, an exponential input to a discrete system will produce an exponential response.

$$y(n) + a_1 y(n-1) + a_2 y(n-2) + \cdots + a_p y(n-p) = b_0 x(n) + b_1 x(n-1) + \cdots + b_q x(n-q)$$

Let us now substitute the exponential sequence $x(n) = zn$. Here z is a complex number, which may be expressed in the usual complex number forms wherever that becomes useful in the analysis. Let us assume that the response is of the form $y(n) = H(z)zn$. Substitution into the difference equation then produces

$$\left(1 + a_1 z^{-1} + a_2 z^{-2} + \cdots + a_p z^{-p}\right)H(z) = \left(b_0 + b_1 z^{-1} + \cdots + b_q z^{-q}\right)$$

This solved for the transferred function given by:

$$H(z) = \frac{b_0 + b_1 z^{-1} + \cdots + b_q z^{-q}}{1 + a_1 z^{-1} + a_2 z^{-2} + \cdots + a_p z^{-p}} \tag{2.31}$$

The close relationship between the system function and the difference equation is evident. The system function is defined wherever the denominator is not zero. These locations are called the "poles" of the system function.

2.6.2.1 Delay Operator

Suppose that the z-transform of $x(n)$ is $X(z)$. Then, the z-transform of x $(n - k)$ is $X(z)z - k$. This is shown by the substitution of $x(n - k)$. We can, therefore, refer to z^{-1} as the delay operator.

2.6.2.2 Convolution Property

Let $x_1(n)$ and $x_2(n)$ be two sequences with z-transform relationships

$$x_1(n) \Leftrightarrow X_1(z)$$

$$x_2(n) \Leftrightarrow X_2(z)$$

Let $x(n) = x_1(n)*x_2(n)$ be the convolution operation. Then the z-transform is determined by the product $X(z) = X_1(z)X_2(z)$.

2.6.2.3 Impulse Function

The function $\delta(n)$ is defined as the z-transform $\Delta(z) = 1$, as can be easily established from the definition (2.7).

$$x(z) = \sum_{-\infty}^{\infty} x(n)z^{-n} \tag{2.32}$$

2.6.2.4 Impulse Response

The impulse response of a discrete linear time-invariant system is called $h(n)$.

It is the response of the system to an input $\delta(n)$. The output for any other input sequence $x(n)$ is given by the convolution

$$y(n) = \sum_{m=-\infty}^{\infty} x(m)h(n-m) = \sum_{m=-\infty}^{\infty} h(m)x(n-m) \tag{2.33}$$

Upon using the convolution property, we find that:

$$Y(z) = X(z)H(z) \tag{2.34}$$

In particular, if the input is an impulse then $X(z) = 1$ and the z-transform of the response is just $H(z)$. Therefore, $h(n)$ and $H(z)$ are a transform pair

$$h(n) \Leftrightarrow H(z) \tag{2.35}$$

2.6.2.5 Frequency Response

Just as in the continuous case, we can use an exponential input with a discrete system to determine its frequency response. Let $x(n) = ej\omega n$ be an input sequence. The output from (2.8) is

$$y(n) = \sum_{m=-\infty}^{\infty} h(m)e^{j(n-m)} = e^{jwn} \sum_{m=-\infty}^{\infty} h(m)e^{-jwm} \tag{2.36}$$

The summation is just $H(z)$ with $z = ej\omega$. Hence, an exponential input sequence produces the exponential output sequence

$$y(n) = H\left(e^{j\omega}\right)e^{j\omega n}$$

where

$$H(e^{jw}) = \sum_{m=-\infty}^{\infty} h(m)e^{-jwm} \tag{2.37}$$

The system response $H(e^{j\omega})$ provides the same kind of information about a discrete system that $H(j\omega)$ provides about a continuous mode. The position of the frequency variable in the exponent means that the function is periodic with period 2π.

The frequency response of the system is

$$H\left(e^{jw}\right) = \sum_{m=-\infty}^{\infty} (-a)me^{-jw}m = \sum_{m=-\infty}^{\infty} \left(-a^{-jw}\right)^{m} = \frac{1}{1+ae^{jw}} \tag{2.38}$$

The gain and phase shift of the system at frequency ω is equal to the magnitude and phase angle of $H(ej\omega)$. These are

$$\left|H\left(e^{j\omega}\right)\right| = \frac{1}{\sqrt{(1+a\cos\omega)^2 + a^2\sin^2\omega}} \tag{2.39}$$

$$\phi(e^{jw}) = -\tan^{-1}\frac{a\sin\omega}{1+a\cos\omega} \tag{2.40}$$

The periodic character of both functions is evident since all of the terms in each expression are periodic.

2.7   SYSTEM OUTPUT

The response of a system can be calculated in several ways. The difference equation can be implemented directly, transforms can be used, and the output can be calculated by the general implementation of a digital filter. The program filter is one useful computational implementation. An example of such a computation is provided by sigdemo 3 and sigdemo 5. Finally, sigdemo 4 shows how the spectrum estimate can be improved by averaging many repetitions of power spectrum computations.

2.7.1   CAUSALITY

The output signal depends on the present and previous values of the input.

Forward difference system (not causal(+)).

$$y[n] = x[n+1] + x[n+2] + x[n+3]$$

Backward difference system (casual(−)).

$$y[n] = x[n] + x[n-1] + x[n-2]$$

2.7.2   STABILITY

Stability is one which produces a finite or bounded output in response to a bounded input.

$$S = \sum_{k=-\infty}^{\infty} |h(k)| < \infty$$

This is stable if S is finite.

2.7.3 INVERTIBILITY

If a digital processor with input $x[n]$ gives an output $y[n]$ then its inverse would produce $x[n]$, if fed with $y[n]$.

$$y[n] = m\, x[n] \rightarrow x[n] = \frac{1}{m}\, y[n] \text{ invertability.}$$

$$y[n] = \left(x[n]\right)^3 \rightarrow x[n] = \sqrt[3]{y[n]} \text{ no-invertibility.}$$

2.7.4 MEMORY

A processor possesses memory if its present output $y[n]$ contains a storage or delay element and the output depends upon one or more previous input value as $x[n-1]$, $x[n-2]$,...

Example 2.14

The DSP system has $x[n]$ input and $y[n]$ output, and determines which of the properties, linearity, time-invariance, causality, stability, invertibility, and memory, are possessed by systems defined by the following sequences:

a) $y[n] = x[n] - x[n-1]$
b) $y[n] = y[n-1] + x[n+1]$
c) $y[n] = n\, x[n]$
d) $y[n] = \cos(x[n])$

SOLUTION

a) Linear, time-invariant, causal, stable, invertible, and memory.
b) Linear, time-invariant, noncausal, unstable, invertible, and memory.
c) Linear, time-variant, causal, stable, invertible, and no-memory.
d) Non-linear, time-invariant, causal, stable, not-invertible, and no-memory.

Example 2.15

Compute the system output

$$y(n) = 1.5\, y(n-2) + 2\, x(n-1) + 2.2\, x(n)$$

for the first four samples using the following initial conditions:

(a) Initial conditions: $y(-2) = 0.5, y(-1) = 0, x(-1) = -1$, and input $x(n) = (0.75)^n\, u(n)$
(b) Zero initial conditions: $y(-2) = y(-1) = 0$, and $x(-1) = 0$, and input $x(n) = (0.75)^n\, u(n)$

SOLUTION

(a) For $n = 0$, and using the initial conditions, we obtain the input and output as
$x(0) = (0.75)^0\, u(0) = 1$

$$y(0) = 1.5\, y(-2) + 2\, x(-1) + 2.2\, x(0)$$

$$= 1.5 \times 0.5 - 2 \times (-1) + 2.2 \times 1 = 4.95$$

For $n = 1$, and using the initial conditions, we obtain the input and output as

$$x(1) = (0.75)^1\, u(1) = 0.75$$

$$y(1) = 1.5\,y(-1) + 2\,x(0) + 2.2\,x(1)$$

$$= 1.5 \times 0 - 2 \times 1 + 2.2 \times 0.75 = -0.35$$

$n = 2$,

$$x(2) = (0.75)^2\,u(2) = 0.5625$$

$$y(2) = 1.5\,y(0) + 2\,x(1) + 2.2\,x(2)$$

$$= 1.5 \times (4.95) - 2 \times 0.75 + 2.2 \times 0.5625 = 7.1625$$

$n = 3$,

$$x(3) = (0.75)^3\,u(3) = 0.4218$$

$$y(3) = 1.5\,y(1) + 2\,x(2) + 2.2\,x(3)$$

$$= 1.5 \times (-0.35) - 2 \times 0.5625 + 2.2 \times 0.4218$$

$$= -0.722$$

It can be seen that the further value of the output can be obtained recursively.

(b) Setting $n = 0$, and using the initial conditions, we obtain the input and output as

$$x(0) = (0.75)^0\,u(0) = 1$$

$$y(0) = 1.5\,y(-2) + 2\,x(-1) + 2.2\,x(0)$$

$$= 1.5 \times 0 - 2 \times 0 + 2.2 \times 1 = 2.2$$

For $n = 1$, and using the initial conditions, we obtain the input and output as

$$x(1) = (0.75)^1\,u(1) = 0.75$$

$$y(1) = 1.5\,y(-1) + 2\,x(0) + 2.2\,x(1)$$

$$= 1.5 \times 0 - 2 \times 1 + 2.2 \times 0.75 = -0.35$$

For $n = 2$, and using the past values of the input and output,

$$x(2) = (0.75)^2\,u(2) = 0.5625$$

$$y(2) = 1.5\,y(0) + 2\,x(1) + 2.2\,x(2)$$

$$= 1.5 \times 2.2 - 2 \times 0.75 + 2.2 \times 0.5625 = 3.0375$$

For $n = 3$,

$$x(3) = (0.75)^3\,u(3) = 0.4218$$

$$y(3) = 1.5\,y(1) + 2\,x(2) + 2.2\,x(3)$$

$$= 1.5 \times (-0.35) - 2 \times 0.5625 + 2.2 \times 0.4218 = -0.722$$

The further value of the output can be obtained recursively.

Example 2.16

Compute the DSP system output

$$y(n) = 5\,x(n) + 2\,x(n-1) - 1.5\,y(n-1) - 2.5\,y(n-2)$$

with the initial conditions $y(-2) = 1$, $y(-1) = 0$, $x(-1) = -1$, and input $x(n) = (0.5)^n\,u(n)$.

(a) Compute the system response $y(n)$ for 25 samples using MATLAB.

SOLUTION

A MATLAB program to compute the system response for 20 samples is given below along with the corresponding output shown in graphical form.

```
Clc;
Clear all;
Close all;
xi = [0 -1];
yi = [1 0];
n = 0:1:24;
x = (0.5).^n;
x = [xi x];
y = [];
y = [yi y];
for k = 3:1:27
  r = 5*x(k-2)+2*x(k-1)-1.5*y(k-1)-2.5*y(k-2);
  y = [y r];
end
  subplot(2,1,1), stem(n,x(3:27),'filled','LineWidth',2), grid on
xlabel('n'); ylabel('x(n)');
subplot(2,1,2), stem(n,y(3:27),'filled','LineWidth',2), grid on
xlabel('n'); ylabel('y(n)');
```

Figure 2.14 shows plots of the input and system output for Example 2.16.

PROBLEMS

2.1 Write a program to determine the even and odd parts of the signal $x(t)$ given by

$$x(t) = \begin{cases} A e^{-at} & t > 0 \\ 0 & t < 0 \end{cases}$$

Then, run the program and display $x(t)$, and the even and odd parts.
Assume $\alpha > 0, A > 0$.

2.2 Write a program to find the even and odd parts of the following sequences:
   a) $x_1(n) = u(n)$
   b) $x_2(n) = an\, u(n), a > 0$.
   Then, use this program to display $x_1(n)$, $x_2(n)$, and the even and odd parts of each sequence.

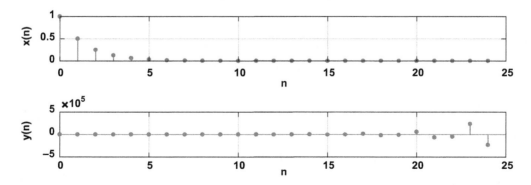

FIGURE 2.14   Plots of the input and system output for Example 2.16.

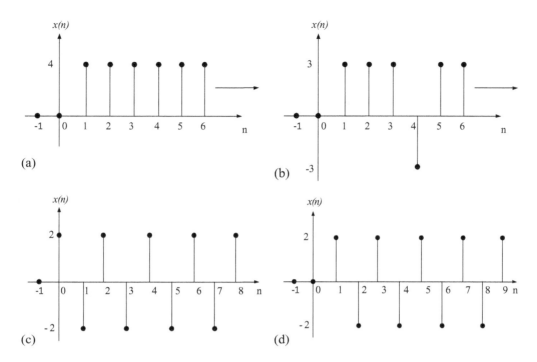

FIGURE 2.15 Sample sequences for Problem 2.3.

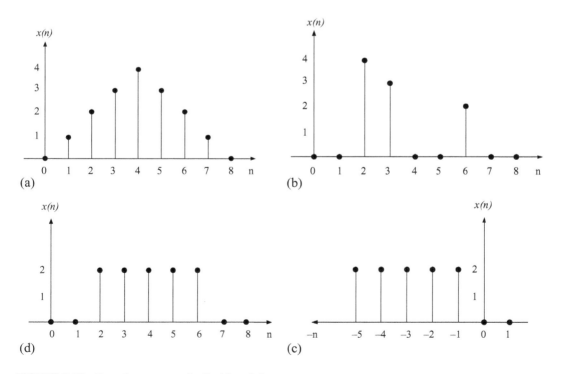

FIGURE 2.16 Sample sequences for Problem 2.5.

2.3   Find the mathematical expressions for the following signals shown in Figure 2.15

2.4   Sketch the following signals

    a $x(n) = -u(n-3)$

    b $y(n) = u(n+2) + 2\delta(n-1)$

    c $g(n) = r(n-1) - 2r(n-4)$

2.5   Find the mathematical expressions for the following signals below (Figure 2.16)

2.6   Sketch the following sequences:

    a. $x(n) = \delta(n+2) - 2\delta(n+1) - 1.5\delta(n) + 2\delta(n-1) - 3\delta(n-3)$

    b. $x(n) = 2\delta(n+1) - 4\delta(n-1) + 2.5u(n-4) + 2.5u(n-6)$

    c. $x(n) = -2\delta(n+1) - 2\delta(n-1) + 4u(n-4)$

    d. $x(n) = \delta(n+3) - 2\delta(n+1) + 2u(n-2) + 4u(n-3)$

2.7   Determine which of the following is a linear system:

    a. $y(n) = 2x(n) + 2x(n-2) - 2x^2(n)$

    b. $y(n) = x(n-2) + x(n-1) - x(n) + 4x(n)$

    c. $y(n) = 2.5x^3(n-1) - 2x(n-2) + 2x(n-1) - 3x(n)$

2.8   Given the following linear systems, find which one is time-invariant:

    a. $y(n) = -2\,x(n-2)$

    b. $y(n) = 3\,x(n-3) + 2\,x(n-2)$

    c. $y(n) = 4\,x(n-6) + 2\,x(n-3)$

    d. $y(n) = 4\,x(n^2)$

2.9   Determine which of the following linear systems is causal:

    a. $y(n) = 0.5\,x(n) + 100\,x(n-2) - 20\,x(n-4)$

    b. $y(n) = x(n+4) + 0.5\,x(n) - 2\,x(n-2)$

    c. $y(n) = 0.5\,y(n-2) + 100\,x(n-2) - 20\,x(n-3)$

    d. $y(n) = 0.5\,y(n-1) - 1 + 100\,x(n-1) - 20\,x(n-2)$

2.10 Determine causality for each of the following linear systems:

    a. $y(n) = 2.5\,x(n) + 6\,x(n-2) - 2.1\,x(n-1)$

    b. $y(n) = y(n+2) - 2.5\,x(n-1)$

    c. $y(n) = y(n-1) + 1.5\,x(n+2)$

2.11 Find the unit-impulse response for each of the following linear systems:

    a. $y(n) = 1.5\,x(n) - 2.5\,x(n-1);$ for $n \geq 0, x(-2) = 0, x(-1) = 0$

    b. $y(n) = 1.75\,y(n-1) + 2\,x(n);$ for $n \geq 0, y(-1) = 0,$

    c. $y(n) = -0.8\,y(n-1) - x(n-1) + 2\,x(n);$ for $n \geq 0, x(-1) = 0, y(-1) = 0$

2.12 Determine stability for the following linear system:

$$y(n) = 0.5\,x(n-1) + 2\,x(n-2) - 4\,x(n-3)$$

$x(n)$ and $y(n)$ are the input and output signals of a digital processor. Determine how the following properties are exhibited by each of the systems defined below: linearity, time-invariance, causality, stability, and memory:

    a) $y(n) = -x(4-n)$

    b) $y(n) = x(n) + x(n-1) - 2\,x(n-2)$

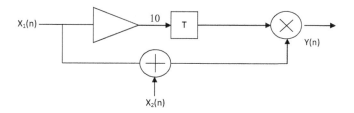

FIGURE 2.17   System for Problem 2.14.

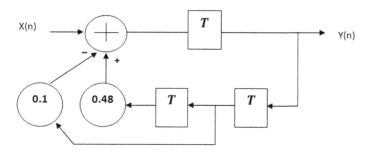

FIGURE 2.18   System for Problem 2.15.

c) $y(n) = n \cdot x(n)$

2.13 Draw a block diagram for a digital processor with the following recurrence formula. Distinguish clearly between its nonrecursive and recursive memory.

a) $y(n) = 1.5\,y(n-1) - 0.5\,y(n-2) + 0.75\,x(n) - 0.4\,x(n-2)$

b) $y(n) = 1.35\,y(n-1) - 0.7\,y(n-3) + 1.75\,x(n) + 2.5\,x(n-1)$

c) $y(n) = 0.5\,y(n-1) - 0.7\,y(n-1) + 0.9\,x(n) - 1.5\,x(n-1)$

2.14 Find the output response for the following system shown in Figure 2.17 if:

$$X_1(n) = 0.1\,\delta(n) + 0.5\,\delta(n-1) + 0.6\,\delta(n-2) + 0.6\,\delta(n-3)$$

$$X_2(n) = \delta(n+1) + \delta(n) + 1.5\,\delta(n-1) - 1.6\,\delta(n-2)$$

2.15 Find the response of the system shown in Figure 2.18 due to a unit step input.

3 Convolution

Characteristics of linear systems are described by the system's impulse response, as administrated by the mathematics of convolution. The important of the impulse response is because of its uses in many practical aspects, as echo suppression in long-distance telephone calls. This chapter describes the properties of one of the most common applications of impulse response called convolution. First, several conventional coevolution methods discussed. Second, it will be presented dealing with cascade and parallel combinations of linear systems. Third, the technique of correlation will introduce. Fourth, a different problem with convolution examined.

3.1 LINEAR CONVOLUTION

A system that is both linear and shift-invariant (time-invariant) is referred to as a *linear shift-invariant* (LSI) system. If $x(n)$ is the response of an LSI system to the unit sample $\delta(n)$, its response to $\delta(n - k)$ will be $h(n - k)$. Therefore, in the superposition the sum is given as:

$$y(n) = \sum_{k=-\infty}^{\infty} x(k)h_k(n) \tag{3.1}$$

then it becomes: $h_k(n) = h(n - k)$
and it follows that

$$y(n) = \sum_{k=-\infty}^{\infty} x(k)h(n - k) \tag{3.2}$$

This equation is known as the convolution sum and written as

$$y(n) = x(n) * h(n) \tag{3.3}$$

Where * indicates the convolution operator. The sequence $h(n)$ referred to as the *unit sample response* or *impulse response* provides a complete characterization of an LSI system. In other words, the response of the system to *any* input $x(n)$ is found once $h(n)$ is known.

3.2 CONVOLUTION PROPERTIES

Convolution is a linear operator and, therefore, has several essential properties, including the commutative, associative, and distributive properties.

3.2.1 COMMUTATIVE PROPERTY

The commutative property states that the order in which two sequences convolved is not essential. Mathematically, the commutative property is

$$x(n) * h(n) = h(n) * x(n) \tag{3.4}$$

From a systems point of view, this property states that a system with a unit sample response $h(n)$ and input $x(n)$ behaves in the same way as a system with a unit sample response $x(n)$ and an input $h(n)$. This illustrated in Figure 3.1(a) (Figure 3.1).

FIGURE 3.1   The commutative property.

$$\left[x(n) * h_1(n)\right] * h_2(n) = x(n) * \left[h_1(n) * h_2(n)\right] \tag{3.5}$$

3.2.2   ASSOCIATIVE PROPERTY

The convolution operator satisfies the associative property, which is

$$h_{eq}(n) = h_1(n) * h_2(n) \tag{3.6}$$

From a systems point of view, the associative property states that if two systems with unit sample responses $h_1(n)$ and $h_2(n)$ are connected in cascade as shown in Figure 3.1(b), an equivalent system is one that has a unit sample response equal to the convolution of $h_1(n)$ and $h_2(n)$ (Figure 3.2):

3.2.3   DISTRIBUTIVE PROPERTY

The distributive property of the convolution operator expresses as

$$x(n) * \left[h_1(n) + h_2(n)\right] = x(n) * h_1(n) + x(n) * h_2(n) \tag{3.7}$$

From a systems point of view, this property asserts that if two systems with unit sample responses $h_1(n)$ and $h_2(n)$ are connected in parallel, as illustrated in Figure 3.4(c), an equivalent system is one that has a unit sample response equal to the sum of $h_1(n)$ and $h_2(n)$ (Figure 3.3):

$$h_{eq}(n) = h_1(n) + h_2(n) \tag{3.8}$$

3.3   TYPES OF CONVOLUTIONS

Several different approaches are used, and the one that is the easiest will depend upon the form and type of sequences convolved.

1. *Equations Method*
2. *Graphical Method*
3. *Alternative Method*

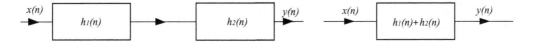

FIGURE 3.2   The associative property.

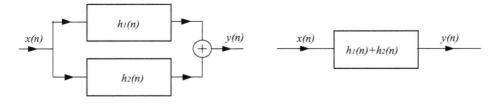

FIGURE 3.3   The distribution property.

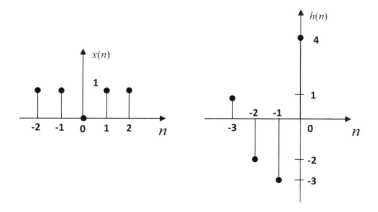

FIGURE 3.4 The convolution sequences.

3.3.1 Equations Method

1) N is the number of samples of the resultant signal from convolution processing for one period

$$N = N_1 + N_2 - 1 \tag{3.9}$$

where N_1 is the number of samples of the first signal,
   N_2 is the number of samples of the second signal, and
   N is the number of samples of the resultant signal.

2) The left and right extremes are found using the left and right extremes of the two sequences to be convolved, i.e.,

$$y_l = x_l + h_l \tag{3.10}$$

$$y_r = x_r + h_r \tag{3.11}$$

Where x_l, h_l, and y_l are the left extremes of the signals x, h, and y, respectively. In a similar manner, x_r, h_r, and y_r are the right extremes of the signals x, h, and y, respectively.

Example 3.1

Find the convolution of the two discrete-time signals which are given below:

$$x(n) = [1, 1, 0, 1, 1], \quad \text{and} \quad h(n) = [1, -2, -3, 4]$$

where $x(0) = 0$, and $h(0) = 4$.

SOLUTION

From the graphs shown in Figure 3.4, we have $x_l = -2$, $x_r = 2$, $h_l = -3$, and $h_r = 0$
   Therefore, the left and right extremes of the convolved signal $y_L(n)$ are

$$y_l = x_l + h_l = -2 + (-3) = -5$$

$$y_r = x_r + h_r = 2 + 0 = 2$$

The number of samples is:

$$N_1 = 5, N_2 = 4 \Rightarrow N = N_1 + N_2 - 1 = 5 + 4 - 1 = 8$$

The convolved signal $y_L(n)$ is expressed as

$$y_L(n) = \sum_{k=-\infty}^{\infty} x(k)h(n-k)$$

when $n = 0$

$$y_L(0) = \sum_{k=-\infty}^{\infty} x(k)h(-k)$$

$$= x(-2)h(2) + x(-1)h(1) + x(0)h(0) + x(1)h(-1) + x(2)h(-2)$$

$$= (1)(0) + (1)(0) + (0)(4) + (1)(-3) + (1)(-2) = -5$$

when $n = 1$

$$y_L(1) = \sum_{k=-\infty}^{\infty} x(k)h(1-k)$$

$$= x(-2)h(3) + x(-1)h(2) + x(0)h(1) + x(1)h(0) + x(2)h(-1)$$

$$= (1)(0) + (1)(0) + (0)(0) + (1)(4) + (1)(-3) = 1$$

when $n = 2$

$$y_L(2) = \sum_{k=-\infty}^{\infty} x(k)h(2-k)$$

$$= x(-2)h(4) + x(-1)h(3) + x(0)h(2) + x(1)h(1) + x(2)h(0)$$

$$= (1)(0) + (1)(0) + (0)(0) + (1)(0) + (1)(4) = 4$$

when $n = -1$

$$y_L(-1) = \sum_{k=-\infty}^{\infty} x(k)h(-1-k)$$

$$= x(-2)h(1) + x(-1)h(0) + x(0)h(-1) + x(1)h(-2) + x(2)h(-3)$$

$$= (1)(0) + (1)(4) + (0)(-3) + (1)(-2) + (1)(1) = 3$$

when $n = -2$

$$y_L(-2) = \sum_{k=-\infty}^{\infty} x(k)h(-2-k)$$

$$= x(-2)h(0) + x(-1)h(-1) + x(0)h(-2) + x(1)h(-3) + x(2)h(-4)$$

$$= (1)(4) + (1)(-3) + (0)(-2) + (1)(1) + (1)(0) = 2$$

when $n = -3$

$$y_L(-3) = \sum_{k=-\infty}^{\infty} x(k)h(-3-k)$$

$$= x(-2)h(-1) + x(-1)h(-2) + x(0)h(-3) + x(1)h(-4) + x(2)h(-5)$$

$$= (1)(-3) + (1)(-2) + (0)(1) + (1)(0) + (1)(0) = -5$$

when $n = -4$

$$y_L(-4) = \sum_{k=-\infty}^{\infty} x(k)h(-4-k)$$

$$= x(-2)h(-2) + x(-1)h(-3) + x(0)h(-4) + x(1)h(-5) + x(2)h(-6)$$

$$= (1)(-2) + (1)(1) + (0)(0) + (1)(0) + (1)(0) = -1$$

when $n = -5$ (Figure 3.5)

$$y_L(-5) = \sum_{k=-\infty}^{\infty} x(k)h(-5-k)$$

$$= x(-2)h(-3) + x(-1)h(-4) + x(0)h(-5) + x(1)h(-6) + x(2)h(-7)$$

$$= (1)(1) + (1)(0) + (0)(0) + (1)(0) + (1)(0) = 1$$

Example 3.2

Find the response of the filter with an impulse response $h(n) = [1,2,4]$ to the input sequence $x(n) = [1,2]$.

SOLUTION

The type of convolution is a *linear convolution*.
Here, $N_1 = 2$ and $N_2 = 3$. Then $N = N_1 + N_2 - 1 = 4$.

$$y_L(n) = \sum_{k=-\infty}^{\infty} x(k)h(n-k)$$

Therefore,

$$y_L(0) = \sum_{k=-\infty}^{\infty} x(k)h(-k) = x(0)h(0) + x(1)h(-1) = (1)(1) + (2)(0) = 1$$

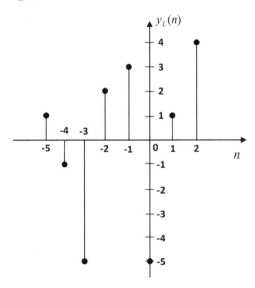

FIGURE 3.5 The convoluted sequence.

$$y_L(1) = \sum_{k=-\infty}^{\infty} x(k)h(1-k) = x(0)h(1) + x(1)h(0) = (1)(2) + (2)(1) = 4$$

$$y_L(2) = \sum_{k=-\infty}^{\infty} x(k)h(2-k) = x(0)h(2) + x(1)h(1) = (1)(4) + (2)(2) = 8$$

$$y_L(3) = \sum_{k=-\infty}^{\infty} x(k)h(3-k) = x(0)h(3) + x(1)h(2) = (1)(0) + (2)(4) = 8$$

$$\Rightarrow y_L(n) = [1, 4, 8, 8]$$

Example 3.3

Determine the linear convolution of the two finite duration sequences given below:

$$x(n) = \begin{cases} 1 & -1 \le n \le 1 \\ 0 & \text{otherwise} \end{cases} \quad \text{and} \quad h(n) = \begin{cases} 1 & -1 \le n \le 1 \\ 0 & \text{otherwise} \end{cases}$$

SOLUTION

We know that the convolution of the two sequences is expressed as

$$y_L(n) = \sum_{k=-\infty}^{\infty} x(k)h(n-k) = \sum_{k=-\infty}^{\infty} x(n-k)h(k)$$

First plot the given sequences (Figure 3.6).
To find the convolution, we have

$$n = 0 \Rightarrow y_L(0) = \sum_{k=-\infty}^{\infty} x(k)h(-k)$$

$$= x(-1)h(1) + x(0)h(0) + x(1)h(-1)$$

$$= (1)(1) + (1)(1) + (1)(1) = 3$$

when

$$n = 1 \Rightarrow y_L(1) = \sum_{k=-\infty}^{\infty} x(k)h(1-k)$$

$$= x(-1)h(2) + x(0)h(1) + x(1)h(0)$$

$$= (1)(0) + (1)(1) + (1)(1) = 2$$

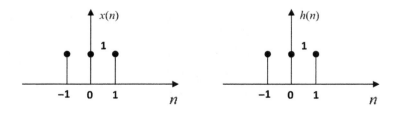

FIGURE 3.6   The convolution sequences.

when

$$n = 2 \Rightarrow y_L(2) = \sum_{k=-\infty}^{\infty} x(k)h(2-k)$$

$$= x(-1)h(3) + x(0)h(2) + x(1)h(1)$$

$$= (1)(0) + (1)(0) + (1)(1) = 1$$

when

$$n = -1 \Rightarrow y_L(-1) = \sum_{k=-\infty}^{\infty} x(k)h(-1-k)$$

$$= x(-1)h(0) + x(0)h(-1) + x(1)h(-2)$$

$$= (1)(1) + (1)(1) + (1)(0) = 2$$

when (Figure 3.7)

$$n = -2 \Rightarrow y_L(-2) = \sum_{k=-\infty}^{\infty} x(k)h(-2-k)$$

$$= x(-1)h(-1) + x(0)h(-2) + x(1)h(-3)$$

$$= (1)(1) + (1)(0) + (1)(0) = 1$$

3.3.1.1 Convolution of Two Sequences in MATLAB

To write a MATLAB program to find the convolution of two sequences follow the below.

```
clc;
clear all;
close all;
n=0:8;
x1=1;
x2=0;
y1=x1.*(n>=0 & n<=2)+x2.*(n>=2 & n<=8);
subplot(2,2,1);
stem(n,y1);
axis([0 8 0 1.5]);
xlabel('time n ---->');
```

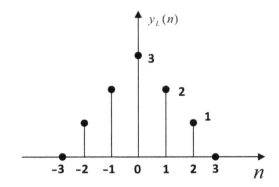

FIGURE 3.7 The convoluted sequence.

```
ylabel('amplitude---->');
title('the sequence y1[n]')
y2=x1.*(n>=0 & n<=4)+x2.*(n>=4 & n<=8);
subplot(2,2,2);
stem(n,y2);
axis([0 8 0 1.5]);
xlabel('time n ---->');
ylabel('amplitude---->');
title('the sequence y2[n]')
y=conv(y1,y2);
L=length(y);
n=0:L-1;
subplot(2,2,[3,4]);
stem(n,y);
axis([0 10 0 4]);
xlabel('time n ---->');
ylabel('amplitude---->');
title('the convolution sequence of y1[n]&y2[n]');
```

Figure 3.8 shows the convolution and convoluted sequences obtained from running the MATLAB program.

3.3.2 GRAPHICAL METHOD

The steps for finding out the convolution sum are as follows:

1) **Plotting**: Plot both sequences, as a function of k.
2) **Folding**: Fold the signal $x_2(k)$ about the origin, meaning $x_2(n)$ becomes $x_2(-n)$.
3) **Shifting**: Shift $x_2(-k)$ to the right by, n_0 it is positive or shift $x_2(-k)$ to the left by n_0 if n_0 it is negative to obtain $x_2(n_0 - k)$.
4) **Multiplication**: Multiply the product sequence $x_2(k)$ by $x_2(n_0 - k)$ to obtain the product sequence.
5) **Summation**: Sum all the values of the product sequence to obtain the value of the output at a time $n = n_0$.

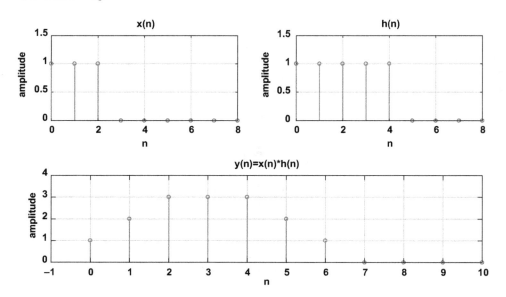

FIGURE 3.8 The convolution and convoluted sequences.

Example 3.4

The previous example can also be solved using the graphical method below.
We know that (Figure 3.9)

$$y_L(n) = \sum_{k=-\infty}^{\infty} x(k)h(n-k)$$

Alternative Method: Using the matrix representation, the linear convolution of the given two sequences is determined as in Table 3.1
So the result of convolution is given as $y_L(n) = [1, 4, 8, 8]$.

3.3.3   TABULAR METHOD

Example 3.5

Find the convolution between

$$x(n) = \begin{bmatrix} 1, 2, 3, 4, 5 \end{bmatrix} \text{ and } h(n) = \begin{bmatrix} 5, 4, 3, 2, 1 \end{bmatrix}$$

SOLUTION

N1 = 5, N2 = 5, N1 + N2 = 5 + 5 = 10, N1 + N2 = 1 = 10 − 1 = 9 so the output sequence will be from $y(0)$ to $y(9)$

					1	2	3	4	5		← $x(n)$
$y(0)$	1	2	3	4	5						← Reversed $h(n)$
$y(1)$		1	2	3	4	5					
$y(2)$			1	2	3	4	5				
$y(3)$				1	2	3	4	5			
$y(4)$					1	2	3	4	5		
$y(5)$						1	2	3	4	5	
$y(6)$							1	2	3	4	5
$y(7)$								1	2	3	4 5
$y(8)$									1	2	3 4 5
$y(9)$										1	2 3 4 5

$y(0) = 1*5 = 5$
$y(1) = 1*4 + 2*5 = 14$
$y(2) = 1*3 + 2*4 + 3*5 = 26$
$y(3) = 1*2 + 2*3 + 3*4 + 4*5 = 40$
$y(4) = 1*1 + 2*2 + 3*3 + 4*4 + 5*5 = 55$
$y(5) = 2*1 + 3*2 + 4*3 + 5*4 = 40$
$y(6)= 3*1 + 4*2 + 5*3 = 26$
$y(7) = 4*1 + 5*2 = 14$
$y(8) = 5*1 = 5$
$y(9) = 0$

Example 3.5
Convolve $x(n) = (0.5)^n u(n)$
With the ramp sequence $h(n) = u(n)$
The convolution of $x(n)$ with $h(n)$ is given by

$$y(n) = x(n) * h(n) = \sum_{k=-\infty}^{\infty} x(k)h(n-k)$$

$$= \sum_{k=-\infty}^{\infty} \left[(0.5)^k u(k) \right] \left[u(n-k) \right]$$

$$y_L(n) = \sum_{k=-\infty}^{\infty} x(k)h(n-k)$$

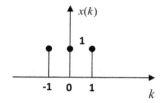

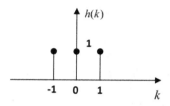

when $n = 0$

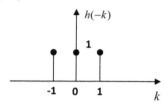

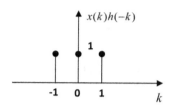

$$y_L(0) = \sum_{k=-\infty}^{\infty} x(k)h(-k) = 3$$

when $n = 1$

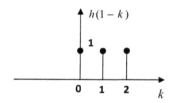

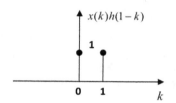

$$y_L(1) = \sum_{k=-\infty}^{\infty} x(k)h(1-k) = 2$$

when $n = 2$

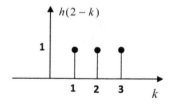

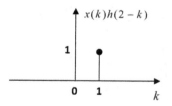

FIGURE 3.9   The convolution and convoluted sequences of Example 3.4.

$u(k) = 0$ for $k < 0$, and $u(n - k) = 0$ for $k > n$, so

$$y(n) = x(n) * h(n) = \sum_{k=0}^{n} [(0.5)^k] \quad n \geq 0$$

$$y(n) = x(n) * h(n) = \sum_{k=0}^{n} (0.5)^k \quad n \geq 0$$

$$y_L(2) = \sum_{k=-\infty}^{\infty} x(k)h(2-k) = 1$$

when $n = 3$

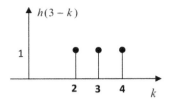

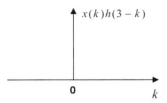

$$y_L(3) = \sum_{k=-\infty}^{\infty} x(k)h(3-k) = 0$$

when $n = -1$

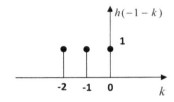

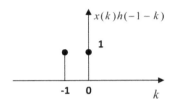

$$y_L(-1) = \sum_{k=-\infty}^{\infty} x(k)h(-1-k) = 2$$

when $n = -2$

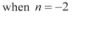
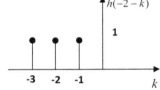

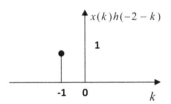

FIGURE 3.9   (Continued)

Using series given in the table, we have

$$y(n) = x(n) * h(n) = \frac{1-(0.5)^{n+1}}{1-0.5}$$

$$= 2\left[1-(0.5)^{n+1}\right] \quad n \geq 0$$

Example 3.6

Convolve $x(n) = (0.8)^n u(n)$ with the ramp sequence $h(n) = n\, u(n)$.
    The convolution of $x(n)$ with $h(n)$ is given by

$$y(n) = x(n) * h(n) = \sum_{k=-\infty}^{\infty} x(k)h(n-k)$$

$$= \sum_{k=-\infty}^{\infty} [(0.8)^k u(k)][(n-k)u(n-k)]$$

$$y_L(-2) = \sum_{k=-\infty}^{\infty} x(k)h(-2-k) = 1$$

when $n = -3$

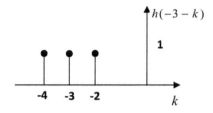

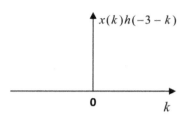

$$y(-3) = \sum_{k=-\infty}^{\infty} x(k)h(-3-k) = 0$$

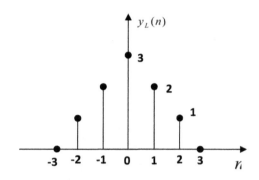

FIGURE 3.9   (Continued)

TABLE 3.1
Convolution of Two Sequences

x(n) \ h(n)	1	2	4
1	1	2	4
2	2	4	8

$u(k) = 0$ for $k < 0$, and $u(n - k) = 0$ for $k > n$, so

$$y(n) = x(n) * h(n) = \sum_{k=0}^{n} [(n-k)(0.8)^k] \quad n \geq 0$$

$$y(n) = x(n) * h(n) = n\sum_{k=0}^{n} (0.8)^k - \sum_{k=0}^{n} k(0.8)^k \quad n \geq 0$$

Using the series given in the table, we have

$$y(n) = x(n) * h(n)$$

$$= n\frac{1-(0.8)^{n+1}}{1-0.8} - \frac{n(0.8)^{n+2}-(n+1)(0.8)^{n-1}+0.8}{(1-0.8)^2}$$

$$= 5n\left[1-(0.8)^{n+1}\right] - 25\left[n(0.8)^{n+2}-(n+1)(0.8)^{n-1}+0.8\right] \quad n \geq 0$$

$$= \left[5n+11.25n(0.8)^n+31.25(0.8)^n+20\right]u(n) \quad n \geq 0$$

Example 3.7

Convolve $x(n) = (0.7)^n \, u(n)$ with the ramp sequence $h(n) = (0.5)^n \left[u(n) - u(n-10)\right]$.

SOLUTION

$$y(n) = x(n) * h(n) = \sum_{k=-\infty}^{\infty} x(k)h(n-k)$$

For $n < 0$, $y(n) = 0$.

For $0 \leq n \leq 10$,

$$y(n) = x(n) * h(n) = \sum_{k=-\infty}^{\infty}(0.5)^k\left[u(n)-u(n-10)\right](0.7)^{n-k}u(n-k)$$

$$= \sum_{k=0}^{10}(0.5)^k(0.7)^{n-k}u(n-k)$$

$u(n-k) = 1$ for $k \leq n$, so

$$y(n) = x(n) * h(n) = \sum_{k=0}^{10}(0.5)^k(0.7)^{n-k}$$

$$y(n) = \sum_{k=0}^{10}(0.5)^k(0.7)^n(0.7)^{-k}$$

$$= (0.7)^n\sum_{k=0}^{10}(0.5)^k(0.7)^{-k}$$

$$= (0.7)^n\sum_{k=0}^{10}\left(\frac{5}{7}\right)^k$$

$$= (0.7)^n\frac{1-\left(\dfrac{5}{7}\right)^{n+1}}{1-\left(\dfrac{5}{7}\right)}$$

$$y(n) = 3.5(0.7)^n\left[1-\left(\frac{5}{7}\right)^{n+1}\right] \quad 0 \leq n \leq 10$$

For $n \geq 10$, $u(n-k) = 1$ for all k in the range $0 \leq k \leq 10$, therefore,

$$y(n) = 3.5(0.7)^n\left[1-\left(\frac{5}{7}\right)^{11}\right] \quad n \geq 10$$

$$y(n) = 3.4135(0.7)^n \quad n \geq 10$$

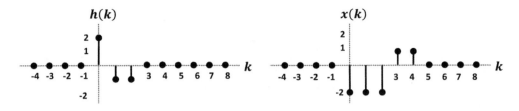

FIGURE 3.10 The convolution sequences of Example 3.8.

Example 3.8

$$y(n) = \sum_{k=-\infty}^{\infty} h(k)x(n-k)$$

a. Using the graphical method;
b. Using the tabular method.

SOLUTION

(a) Graphical method: Sketches of $x(k)$ and $h(k)$ are given in the following (Figures 3.10 and 3.11).
And for $n \geq 7$, $y(n) = 0$.

(b) Tabular method

k	−4	−3	−2	−1	0	1	2	3	4	5	6	7	
$x(k)$					−2	−2	−2	1	1				
$h(k)$					2	−1	−1						
$x(-k)$	1	1	−2	−2	−2								$y(0) = 2 \times -2 = -4$
$x(1-k)$		1	1	−2	−2	−2							$y(1) = 2 \times -2 + (-1) \times (-2) = -2$
$x(2-k)$			1	1	−2	−2	−2						$y(2) = 2 \times -2 + (-1) \times (-2) + (-1) \times (-2) = 0$
$x(3-k)$				1	1	−2	−2	−2					$y(3) = 2 \times 1 + (-1) \times (-2) + (-1) \times (-2) = 6$
$x(4-k)$					1	1	−2	−2	−2				$y(4) = 2 \times 1 + (-1) \times (1) + (-1) \times (-2) = 3$
$x(5-k)$						1	1	−2	−2	−2			$y(5) = (-1) \times (1) + (-1) \times (1) = -2$
$x(6-k)$							1	1	−2	−2	−2		$y(6) = (-1) \times (1) = -1$
$x(7-k)$								1	1	−2	−2	−2	$y(7) = 0$

PROBLEMS

3.1 Find the convolution of the two discrete-time signals which are given below:

$$x(n) = [0,1,2,3,4,5], \text{ and } h(n) = [-1,4,3,2]$$

where $x(0) = 0$, and $h(0) = 4$.

3.2 Find the convolution of the two discrete-time signals which are given below:

$$x(n) = [-1,1,2,3,4,5], \text{ and } h(n) = [-1,4,3,2]$$

where $x(2) = 3$, and $h(-1) = -1$.

To find $y(0)$, we need the reversed

sequence $x(-k)$.

$y(0)$ = sum of the product of $h(k)$ and $x(-k)$

$\Rightarrow y(0) = 2 \times -2 = -4$

Similarly

$y(1)$ = sum of the product of $h(k)$ and $x(1-k)$

$\Rightarrow y(1) = 2 \times (-2) + (-1) \times (-2) = -2$

$y(2)$ = sum of the product of $h(k)$ and $x(2-k)$

$\Rightarrow y(2) = 2 \times (-2) + (-1) \times (-2)$

$\qquad + (-1) \times (-2) = 0$

$y(3)$ = sum of the product of $h(k)$ and $x(3-k)$

$\Rightarrow y(3) = 2 \times (1) + (-1) \times (-2)$

$\qquad + (-1) \times (-2) = 6$

$y(4)$ = sum of the product of $h(k)$ and $x(4-k)$

$\Rightarrow y(4) = 2 \times (1) + (-1) \times (1)$

$\qquad + (-1) \times (-2) = 3$

$y(5)$ = sum of the product of $h(k)$ and $x(5-k)$

$\Rightarrow y(5) = (-1) \times (1) + (-1) \times (1) = -2$

$y(6)$ = sum of the product of $h(k)$ and $x(6-k)$

$\Rightarrow y(5) = (-1) \times (1) = -1$

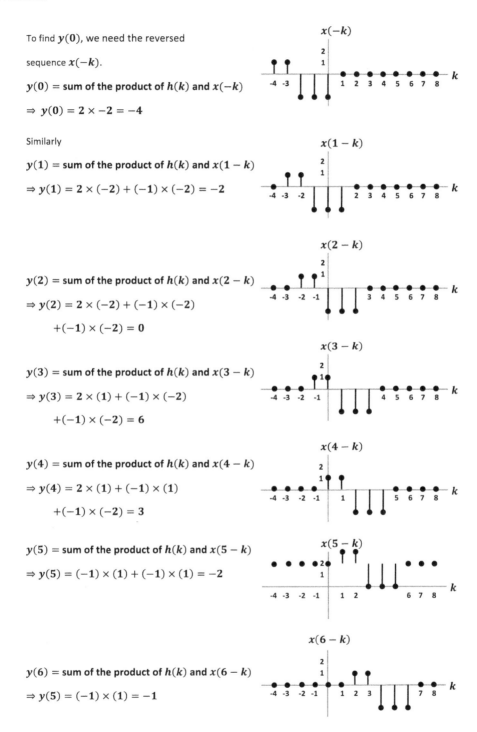

FIGURE 3.11 The convoluted sequence of Example 3.8.

3.3   Find the convolution of the two discrete-time signals which are given below:

$$x(n) = [-1,1,2,5], \text{ and } h(n) = [-1,2,1]$$

where $x(2) = 3$, and $h(0) = -1$

3.4   Write a program to convolve the two digital sequences given by:

$$x_1 = 3r[n+3] - 3r[n-1] - 12u[n-4]$$

$$x_2 = \begin{cases} \exp(n) & \text{for } 0 \le n \le 3 \\ 0 & \text{otherwise} \end{cases}$$

Then, run the program and display the result.

3.5   Write a program to perform the linear convolution of the two sequences below in the time domain.

$$h(n) = \begin{cases} n & \text{for } 0 \le n \le 3 \\ 0 & \text{otherwise} \end{cases}$$

$$x(n) = \begin{cases} |n| & \text{for } -1 \le n \le 1 \\ 0 & \text{otherwise} \end{cases}$$

Use this program, to generate and plot the sequences $h(n)$, $x(n)$, and the output sequence.

3.6   Write a program to compute the output $y(t)$ for a continuous-time LTI system whose impulse response $h(t)$ and the input $x(t)$ are given by:

$$h(t) = e^{-at}u(t), \, x(t) = e^{at}u(-t), \quad a > 0$$

Use this program, to generate and plot the signals $h(t)$, $x(t)$, and $y(t)$.

3.7   Convolve $x(n) = (0.45)^n \, u(n)$
With the ramp sequence $h(n) = n \, u(n)$
The convolution of $x(n)$ with $h(n)$ is given by

$$y(n) = x(n) * h(n) = \sum_{k=-\infty}^{\infty} x(k)h(n-k)$$

3.8   Convolve $x(n) = (0.5)^n \, u(n)$
With the ramp sequence $h(n) = (n-1) \, u(n)$
The convolution of $x(n-2)$ with $h(n)$ is given by

$$y(n) = x(n) * h(n) = \sum_{k=-\infty}^{\infty} x(k)h(n-k)$$

3.9   Convolve $x(n) = (0.75)^n \, u(n)$

With the ramp sequence $h(n) = (0.45)^n \left[u(n) - u(n-5) \right]$

3.10 Convolve $x(n) = (0.9)^n \, u(n)$

With the ramp sequence $h(n) = (0.65)^n \left[u(n) - u(n-6) \right]$

3.11 Find the convolution for the two sequences:

$$x(n) = (0.8)^n \, u(n), \quad h(n) = (0.8)^n \, u(n-1).$$

3.12 Find the convolution for the two of following:

A $X(n) = (0.8)^n u(n), \quad h(n) = (0.8)^n u(n-1).$

B $X(n) = [0,0,0,3,1,2], \quad h(n) = [4,2,3,2].$

C $h(n) = \sin(n * \pi/4) * [u(n) - u(n-4)] * (2n-1)$

$x(n) = \cos(n * \pi/4) * [u(n-1) - u(n-5)]$

3.13 Let $y(n) = x(n) \otimes h(n)$ and $w(n) = x(n-1) \otimes h(n-2)$ where $x(n)$ and $h(n)$ are given as shown below to

find $y(n)$. Express $w(n)$ in terms of $y(n)$ using a z-transformation (Figure 3.12).

3.14 Find the convolution of the two sequences:

$$x(n) = \delta(n-2) - 2\delta(n-4) + 3\delta(n-6)$$

$$h(n) = 2\delta(n+3) + \delta(n) + 2\delta(n-2) + \delta(n-3)$$

3.15 Using the following sequence definitions

$$x(k) = \begin{cases} -1, & k = 0,1,2 \\ 1, & k = 3,4 \\ 0 & \text{elsewhere} \end{cases} \quad \text{and} \quad h(k) = \begin{cases} 1, & k = 0, \\ -1, & k = 1,2 \\ 0 & \text{elsewhere} \end{cases}$$

evaluate the digital convolution.

3.16 Determine stability for each of the following linear systems:

$$y(n) = \sum_{k=0}^{\infty} 0.72^k x(n-k)$$

$$y(n) = \sum_{k=0}^{\infty} 3^k x(n-k)$$

3.17 Given the sequence

$$h(k) = \begin{cases} 3, & k = 0,1,2 \\ 1, & k = 3,4 \\ 0 & \text{elsewhere} \end{cases}$$

where k is the time index or sample number.
a. Sketch the sequence $h(k)$ and the reverse sequence $h(-k)$.
b. Sketch the shifted sequences $h(-k + 2)$ and $h(-k - 3)$.

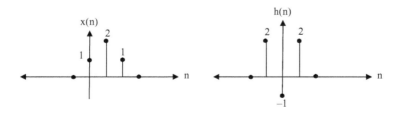

FIGURE 3.12 The convolution sequences of Problem 3.13.

3.18 Using the following sequence definitions

$$h(k) = \begin{cases} 3, & k = 0,1,2 \\ 2, & k = 3,4 \\ 0 & \text{elsewhere} \end{cases} \quad \text{and} \quad x(k) = \begin{cases} 3, & k = 0, \\ 2, & k = 1,2 \\ 0 & \text{elsewhere} \end{cases}$$

evaluate the digital convolution.

$$y(n) = \sum_{k=-\infty}^{\infty} x(k)h(n-k)$$

a. Using the graphical method.
b. Using the tabular method.
c. Applying the convolution formula directly.

3.19 Given the sequence definitions

$$x(k) = \begin{cases} -2 & k = 0,1,2 \\ 1.5 & k = 3,4 \\ 0 & \text{elsewhere} \end{cases} \quad \text{and} \quad h(k) = \begin{cases} 2.5 & k = 0, \\ -1.2 & k = 1,2 \\ 0 & \text{elsewhere} \end{cases}$$

evaluate the digital convolution.

$$y(n) = \sum_{k=-\infty}^{\infty} h(k)x(n-k)$$

a. Using the graphical method.
b. Using the table method.
c. Applying the convolution formula directly.

4 Difference Equations

The signals classify to two types with respect to time. One can think of time as a continuous variable, or one can think of time as a discrete variable. The first case often leads to differential equations. Difference equations relate to differential equations as discrete mathematics relates to continuous mathematics. We will not discuss differential equations in these notes.in this chapter will discuss the difference equation as one application in discrete signals

4.1 DIFFERENCE EQUATIONS AND IMPULSE RESPONSES

Now we study the difference equation and its impulse response.

Format of Difference Equations

A causal, linear, time-invariant system can be described by a difference equation with the following general form:

$$y(n) + b_1 \, y(n-1) + \cdots + b_N \, y(n-N) = a_0 \, x(n) + a_1 \, x(n-1) + \cdots + a_M \, x(n-M) \tag{4.1}$$

where $b_1, \ldots, b_N$ and $a_0, a_1, \ldots, aM$ are the coefficients of the difference equation. The equation above can further be written as:

$$y(n) = -\sum_{j=1}^{N} b_j \, y(n-j) + \sum_{i=0}^{M} a_i \, x(n-i) \tag{4.2}$$

Notice that $y(n)$ is the current output, which depends on the past output samples $y(n-1), \ldots, y(n-N)$, the current input sample $x(n)$, and the past input samples, $x(n-1), \ldots, x(n-N)$.

For an LTI system that is described by a difference equation, the unit sample response, $h(n)$, is found by solving the differential equation for $x(n) = \delta(n)$ assuming initial rest.

For a nonrecursive system, b_j's $= 0$, the difference equation becomes:

$$y(n) = \sum_{i=0}^{M} a_i \, x(n-i) \tag{4.3}$$

And the output is simply a weighted sum of the current and past input values. As a result, the unit sample response is

$$h(n) = \sum_{i=0}^{M} a_i \, \delta(n-i) \tag{4.4}$$

Thus, $h(n)$ is finite in length, and the system is referred to as a finite-length impulse response (FIR) system. However, if b_j's $\neq 0$, the unit sample response is, in general, infinite in length and the system is referred to as an infinite-length impulse response (IIR) system. For example, if:

$$y(n) = b \, y(n-1) + x(n), \text{ the unit sample response is } h(n) = b^n \, u(n). \tag{4.5}$$

Example 4.1

Given the following difference equation:

$$y(n) = 0.25\, y(n-1) + x(n)$$

identify the nonzero system coefficients.

SOLUTION

By comparing variables that lead to: $a_0 = 1$ and $-b_1 = 0.25 \Rightarrow b_1 = -0.25$

Example 4.2

Given a linear system described by the difference equation:

$$y(n) = x(n) + 0.5\, x(n-1)$$

determine the nonzero system coefficients.

SOLUTION

By comparing variables that lead to: $a_0 = 1$ and $a_1 = 0.5$

4.2 SYSTEM REPRESENTATION USING ITS IMPULSE RESPONSE

A linear time-invariant system can be completely described by its unit-impulse response, which is defined as the system response due to the impulse input $\delta(n)$ with zero initial conditions. With the obtained unit-impulse response $h(n)$, we can represent the linear time-invariant system.

Example 4.3

Given the linear time-invariant system

$$y(n) = 0.5\, x(n) + 0.25\, x(n-1), \text{ with an initial condition } x(-1) = 0$$

determine the unit-impulse response $h(n)$.

SOLUTION

Let $x(n) = \delta(n)$, then

$$h(n) = y(n) = 0.5\, x(n) + 0.25\, x(n-1) = 0.5\, \delta(n) + 0.25\, \delta(n-1)$$

Thus, for this linear system, we have:

$$h(n) = \begin{cases} 0.5 & n = 0 \\ 0.25 & n = 1 \\ 0 & \text{elsewhere} \end{cases}$$

Example 4.4

Given the difference equation

$$y(n) = 0.25\, y(n-1) + x(n) \text{ for } n \geq 0 \text{ and } y(-1) = 0$$

determine the unit-impulse response $h(n)$.

SOLUTION

Let $x(n) = \delta(n)$, then, $h(n) = 0.25\,h(n-1) + \delta(n)$

To solve for $h(n)$, we evaluate

$$h(0) = 0.25\,h(-1) + \delta(0) = 0.25 \times 0 + 1 = 1$$

$$h(1) = 0.25\,h(0) + \delta(1) = 0.25 \times 1 + 0 = 0.25$$

$$h(2) = 0.25\,h(1) + \delta(2) = 0.25 \times 0.5 + 0 = 0.0625$$

With the calculated results, we can predict the impulse response as

$$h(n) = (0.25)^n\, u(n) = \delta(n) + 0.25\,\delta(n-1) + 0.0625\,\delta(n-2) + \cdots$$

4.3 THE METHODS THAT ONE MAY USE TO SOLVE THE DIFFERENCE EQUATIONS

There are several different methods that one may use to solve (which means finding $y(n)$) the differential equations for a general input $x(n)$. These are:

1. The tabulation method.
2. The classical approach.
3. Using z-transforms, which will be discussed later.

Important note: The Solving of the difference equation, its mean, finding a form solution for $y(n)$ for a given $x(n)$

The Tabulation Method

The first is to set up a table of input and output values and evaluate the difference equation for each value of n. This approach would be appropriate if only a few output values needed to be determined.

Example 4.5

Given the linear time-invariant system

$$y(n) = 0.5\,x(n) + 0.25\,x(n-1), \text{ with an initial condition } x(-1) = 0$$

a. Determine the unit-impulse response $h(n)$.
b. Write the output using the obtained impulse response.

SOLUTION

a. From a previous example:

$$h(n) = \begin{cases} 0.5 & n = 0 \\ 0.25 & n = 1 \\ 0 & \text{elsewhere} \end{cases}$$

b. Using the convolution of the equation, $y(n)$ can be rewritten as:

$$y(n) = h(0)\,x(n) + h(1)\,x(n-1)$$

From this result, it is noted that if the difference equation without the past output terms, $y(n-1),\ldots,$ $y(n-N)$, that is, the corresponding coefficients $b_1,\ldots, b_N$, are zeros, the impulse response $h(n)$ has a finite number of terms. We call this a finite impulse response (FIR) system.

Example 4.6

Given the difference equation

$$y(n) = 0.25y(n-1) + x(n) \text{ for } n \geq 0 \text{ and } y(-1) = 0$$

   a. Determine the unit-impulse response $h(n)$.
   b. Write the output using the obtained impulse response.
   c. For a step input $x(n) = u(n)$, verify and compare the output responses for the first three output samples using the difference equation and digital convolution sum.

SOLUTION

   a. From a previous example, the predicted impulse response is:

$$h(n) = (0.25)^n u(n) = \delta(n) + 0.25\,\delta(n-1) + 0.0625\,\delta(n-2) + \cdots$$

   b. The output sequence is a sum of infinite terms expressed as

$$y(n) = h(0)x(n) + h(1)x(n-1) + h(2)x(n-2) + \cdots$$

$$= x(n) + 0.25x(n-1) + 0.0625x(n-2) + \cdots$$

   c. From the difference equation and using the zero initial condition, we have

$$y(n) = 0.25y(n-1) + x(n) \text{ for } n \geq 0 \text{ and } y(-1) = 0$$

$$n = 0, y(0) = 0.25y(-1) + x(0) = u(0) = 1$$

$$n = 1, y(1) = 0.25y(0) + x(1) = 0.25u(0) + u(1) = 1.25$$

$$n = 2, y(2) = 0.25y(1) + x(2) = 0.25 \times 1.25 + u(2) = 1.3125$$

$$\cdots$$

Applying the convolution sum yields:

$$y(n) = x(n) + 0.25x(n-1) + 0.0625x(n-2) + \cdots$$

$$n = 0, y(0) = x(0) + 0.25x(-1) + 0.0625x(-2) + \cdots$$

$$= u(0) + 0.25u(-1) + 0.0625u(-2) + \cdots = 1$$

$$n = 1, y(1) = x(1) + 0.25x(0) + 0.0625x(-1) + \cdots$$

$$= u(1) + 0.25u(0) + 0.0625u(-1) + \cdots = 1.25$$

$$n = 2, y(2) = x(2) + 0.25x(1) + 0.0625x(0) + \cdots$$

$$= u(2) + 0.25u(1) + 0.0625u(0) + \cdots = 1.3125$$

Notice that this impulse response $h(n)$ contains an infinite number of terms in its duration due to the past output term $y(n-1)$. Such a system, as described in the preceding example, is called an infinite impulse response (IIR) system. Note that we cannot have a closed-form expression for $y(n)$. Up to now, this is a problem. But using the classical approach, a closed-form expression for $y(n)$ can be obtained.

4.4   THE CLASSICAL APPROACH

In the classical approach you must find the homogeneous and particular solutions. So for a given difference equation, the general solution is a sum of two parts,

$$y(n) = y_h(n) + y_p(n) \tag{4.5}$$

where $y_h(n)$ is known as the homogeneous solution, and $y_p(n)$ is the particular solution. The homogeneous solution is the response of the system to input $x(n) = 0$. The particular solution is the response of the system to input $x(n)$, assuming zero initial conditions.

The homogeneous solution is found by solving the homogeneous difference equation:

$$y(n) + \sum_{j=1}^{N} b_j \, y(n-j) = 0 \tag{4.6}$$

The solution to the above equation is found by assuming a solution of the form:

$$y_h(n) = z^n \tag{4.7}$$

Substituting this solution, we obtain the polynomial equation:

$$z^n + \sum_{j=1}^{N} b_j \, z^{n-j} = 0 \tag{4.8}$$

The polynomial in braces is called the characteristic polynomial.

For p roots z_j, the general solution to the homogeneous differential equation is:

$$y_h(n) = \sum_{j=1}^{p} A_j \, z_j^n \tag{4.9}$$

where the constants A_j are chosen to satisfy the initial conditions.

For a particular solution, it is necessary to find the sequence $y_p(n)$ that satisfies the differential equation for the given $x(n)$. However, for many of the standard inputs that we are interested in, the solution will have the same form as the input.

Table 4.1 lists the particular solution for some commonly encountered inputs. For example, if $x(n) = a^n u(n)$, the particular solution will be of the form:

$$y_p(n) = C \, a^n \tag{4.10}$$

provided a is not a root of the characteristic equation. The constant C is found by substituting the solution into the differential equation. Note that for $x(n) = a^n \delta(n)$, the particular solution is zero. Because $x(n) = 0$, for $n > 0$, the unit sample only affects the initial condition of $y(n)$.

TABLE 4.1
Particular Solution Functions

$x(n)$	Particular Solution
$C\,u(n)$	C_1
$C\,n$	$C_1 n + C_2$
$C\,a^n$	$C_1\,a^n$
$C\cos(n\omega_o)$	$C_1 \cos(n\omega_o) + C_2 \sin(n\omega_o)$
$C\sin(n\omega_o)$	$C_1 \cos(n\omega_o) + C_2 \sin(n\omega_o)$
$C\,a^n \cos(n\omega_o)$	$C_1 a^n \cos(n\omega_o) + C_2 a^n \sin(n\omega_o)$
$C\,\delta(n)$	None (zero)

Example 4.7

Find the solution to the difference equation:

$$y(n) - 0.5y(n-2) = 2x(n)$$

for $x(n) = u(n)$ assuming initial conditions of $y(-1) = 2$ and $y(-2) = 0$.

SOLUTION

We begin by finding the particular solution. From the Table 4.1, we see that for $x(n) = u(n)$:

$$y_p(n) = C_1$$

Substituting this solution into the differential equation, we find

$$C_1 - 0.5C_1 = 2 \Rightarrow C_1 = \frac{2}{1-0.5} \Rightarrow C_1 = 4$$

To find the homogeneous solution, we set $y_h(n) = z^n$, which gives the characteristic polynomial:

$$z^2 - 0.5 = 0 \Rightarrow (z - 0.707)(z + 0.707) = 0$$

Therefore, the homogeneous solution has the form:

$$y_h(n) = A_1(0.707)^n + A_2(-0.707)^n$$

Thus, the total solution is:

$$y(n) = 4 + A_1(0.707)^n + A_2(-0.707)^n \quad n \geq 0$$

The constants A_1 and A_2 must now be found so that the total solution satisfies the given initial conditions, $y(-1) = 1$ and $y(-2) = 0$. Because the total solution given above only applies for $n \geq 0$, we must derive an equivalent set of initial conditions for $y(0)$ and $y(1)$. Evaluating the equation given in the example at $n = 0$ and $n = 1$. We have:

At $n = 0$

$$y(0) - 0.5y(-2) = x(0) = 1$$

$$y(0) - 0.5 * 0 = x(0) = 1 \rightarrow y(0) = 1$$

$n = 1$

$$y(1) - 0.5y(-1) = x(1) = 1$$

$$y(1) - 0.5 * 1 = x(1) = 1 \rightarrow y(1) = 1.5$$

Substituting these derived initial conditions into the total solution, we have:

$$y(n) = 4 + A_1(0.707)^n + A_2(-0.707)^n$$

$$y(0) = 4 + A_1 + A_2 = 1 \rightarrow A_1 + A_2 = -3$$

$$y(1) = 4 + 0.707A_1 - 0.707A_2 = 1.5 \rightarrow A_1 + A_2 = -2.5$$

Solving for A_1 and A_2 we find:

$$A_1 = -3.268$$

$$A_2 = 0.268$$

Thus, the solution is:

$$y(n) = 4 - 3.268(0.707)^n + 0.268(-0.707)^n \quad n \geq 0$$

Example 4.8

Use MATLAB to solve the following discrete-time system difference equation

$$y(n+1) = -2y(n) + 3y(n-1) + 1.5x(n) + 4x(n-1)$$

compute the value of y at $n = 5$, when the input sequence is $x(n) = [1\ -2\ 3\ -4]$, and the initial conditions are $Y(1) = 1$, $y(2) = 1$.

```
x=[1 -2 3 -4];
y(1)=1;
y(2)=1;
for n=2:4
  y(n+1)=-2*y(n)+3*y(n-1)+1.5*x(n)+4*x(n-1);
end
y(5)
```

Ans 21

Example 4.9

Write a MATLAB program to simulate the following difference equation:

$$8y[n] - 5y[n-1] - 2y[n-2] = x[n] + 3x[n-1]$$

for an input, $x[n] = 2n\ u[n]$ and the initial conditions: $y[-1] = 0$ and $y[0] = 1$.
    Find values of $x[n]$, the input signal, and $y[n]$, the output signal and plot these signals over the range, $-1 = n = 10$.

```
n = 1:10;
a = [8 -5 -2];
b = [1 3];
yi = [1 0];
xi = 1;
zi = filtic(b,a,yi,xi);
y = filter(b,a,2.^n,zi)
```

Ans:
$Y(n)$ = [1.2500   2.2813   4.2383   8.2192   16.1966   32.1777   64.1602   128.1445   256.1304
    512.1176]

Example 4.10

Using MATLAB, draw the output response for the difference equation given by

$$y(k) = 0.1 * y(k-1) + .72 * y(k-2) + 5$$

SOLUTION

```
clc;
clear all;
n=[1:25];
y(1) = 1;
```

```
y(2) = 2;
for k=3:25;
  y(k)=0.1*y(k-1)+.72*y(k-2)+5;
end
stem(n,y,'o')
xlabel('time ---->');
ylabel('amplitude---->');
```

Figure 4.1 shows an output result after running the MATLAB program for Example 4.10.

PROBLEMS

4.1   Given the following difference equation

$$y(n) = 0.25\,y(n-2) + x(n-1) - 0.25\,x(n-2)$$

identify the nonzero system coefficients.

4.2   Given the following difference equation

$$y(n) = 0.5\,y(n-1) + 0.2\,x(n-1)$$

identify the nonzero system coefficients.

4.3   Given a linear system described by the difference equation:

$$y(n) = x(n) + 0.5\,x(n-1) + x(n-2) + 0.75\,x(n-3)$$

determine the nonzero system coefficients.

4.4   Given the linear time-invariant system

$$y(n) = 0.75\,x(n) + 0.25\,x(n-1) + 0.85\,x(n-2),\ \text{with an initial condition} = 0$$

determine the unit-impulse response $h(n)$.

4.5   Given the difference equation

$$y(n) = 0.25\,y(n-1) + 0.25\,x(n)\ \text{ for }\ n \geq 0\ \text{ and }\ y(-1) = 0$$

determine the unit-impulse response $h(n)$.

4.6   Given the difference equation

$$y(n) = 0.25\,y(n-1) + 0.25\,x(n-1)\ \text{ for }\ n \geq 0\ \text{ and }\ y(-1) = 0$$

determine the unit-impulse response $h(n)$.

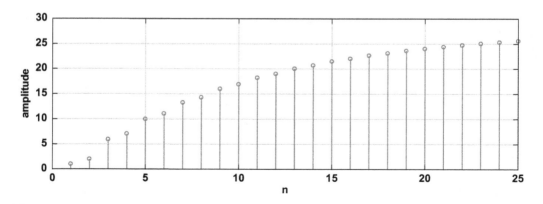

FIGURE 4.1   Output result from the program of Example 4.10.

4.7   Given the linear time-invariant system

$$y(n) = 0.5\,x(n) + 0.75\,x(n-2), \text{ With an initial condition } x(-1) = 0$$

a. Determine the unit-impulse response $h(n)$.
b. Write the output using the obtained impulse response.

4.8   Given the difference equation

$$y(n) = 0.25\,y(n-1) + 0.75\,x(n) \quad \text{for } n \ge 0 \text{ and } y(-1) = 1$$

a. Determine the unit-impulse response $h(n)$.
b. Write the output using the obtained impulse response.
c. Input $x(n) = u(n)$, verify and compare the output responses for the first three output samples using the difference equation and digital convolution sum.

4.9   Given the difference equation

$$y(n) = y(n-1) + 0.75\,x(n) + 0.75\,x(n-1) \quad \text{for } n \ge 0 \text{ and } y(-1) = 1$$

a. Determine the unit-impulse response $h(n)$.
b. Write the output using the obtained impulse response.
c. Input $x(n) = u(n)$, verify and compare the output responses for the first three output samples using the difference equation and digital convolution sum.

4.10 Find the solution to the difference equation:

$$y(n) - 0.5\,y(n-2) = x(n)$$

for $x(n) = u(n)$ assuming initial conditions of $y(-1) = 2$ and $y(-2) = 0$.

4.11 Find the solution to the difference equation

$$y(n) - 0.75\,y(n-2) = a\,x(n)$$

for $x(n) = u(n)$ assuming initial conditions of $y(-1) = 2$ and $y(-2) = 0$.
And a is a positive and constant number.

4.12 Find $H(z)$ for the following difference equations:
   (a) $y[n] = x[n] - 3x[n-1] + 2x[n-2]$
   (b) $y[n] = x[n-1] - 2y[n-1] - 0.5y[n-2]$

4.13 Use MATLAB to recursively determine and plot the system output $y[n]$ for $0 <= n <= 30$ if the system is described by the difference equation

$$y[n] = 0.1y[n-1] + 0.72y[n-2] + 5x[n].$$

The initial conditions are $y[-1] = 1$ and $y[-2] = -1$, and the input is $x[n] = d[n]$ (i.e., the unit-impulse). Present the stem plot with labeled horizontal and vertical axes (i.e., n and $y[n]$ respectively). Based on the stem plot, what can you conclude concerning the stability of the system?

5 Discrete-Time Fourier Series (DTFS)

The representation of periodic signals becomes the discrete-time Fourier series (DTFS), and for aperiodic signals, it becomes the discrete-time Fourier transform(DTFT). The motivation for representing discrete-time signals as a linear combination of complex exponentials is identical in both continuous-time and discrete-time. The complex exponentials are eigenfunctions of linear, time-invariant systems, and consequently, the effect of an LTI system on each of these basic signals is simply the amplitude change. An LTI system is completely considered by a spectrum applies at each frequency. In representing discrete-time periodic signals through the Fourier series, use harmonically related complex exponentials with fundamental frequencies. In this chapter will discuss the discrete-time Fourier transform and its application in digital signal processing.

5.1 DTFS COEFFICIENTS OF PERIODIC DISCRETE SIGNALS

The discrete-time signal $x(n)$ is periodic if for a positive value of N,

$$x(n) = x(n+N) \tag{5.1}$$

Let us look at a process in which we want to estimate the spectrum of a periodic digital signal $x(n)$, sampled at a rate of f_s Hz with the fundamental period $T_0 = NT$, as shown in Figure 5.1, where there are N samples within the duration of the fundamental period, and $T = 1/f_s$ is the sampling period.

According to Fourier series analysis, the coefficients of the Fourier series expansion of a continuous periodic signal $x(t)$ in a complex form is given by:

$$c_k = \frac{1}{T_0} \int_{T_0} x(t) e^{-jk\omega_0 t} dt \quad -\infty < k < \infty \tag{5.2}$$

where k is the number of harmonics corresponding to the harmonic frequency of kf_0, and $\omega_0 = 2\pi/T_0$ and $f_0 = 1/T_0$ are the fundamental frequency in radians per second and in Hz, respectively. To apply this to Equation (5.2), we substitute $T_0 = NT$ with $\omega_0 = 2\pi/T_0$ and approximate the integration over one period using a summation by replacing $dt = T$ and $t = nT$. We obtain:

$$c_k = \frac{1}{N} \sum_{n=0}^{N-1} x(n) e^{-j\frac{2\pi kn}{N}} \quad -\infty < k < \infty \tag{5.3}$$

and, the discrete signal $x(n)$ can be given as:

$$x(n) = \sum_{k=0}^{N-1} c_k e^{j\frac{2\pi kn}{N}} \quad -\infty < n < \infty \tag{5.4}$$

where $\Omega_0 = \frac{2\pi}{N}$.

The value of k varies over any range of N successive integers, and for convenience we indicate this by expressing the limits of the summation as $k = \langle N \rangle$. For example, k could take on the values $k = 0, 1, 2, \ldots, N-1$ or $k = 3, 4, \ldots, N+2$. The two Equations (5.3) and (5.4) could be written in a more general form:

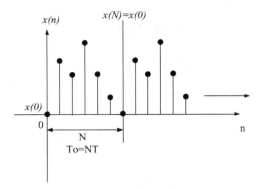

FIGURE 5.1 Periodic discrete signals.

$$c_k = \frac{1}{N} \sum_{n=\langle N \rangle} x(n) e^{-j\frac{2\pi kn}{N}} \quad -\infty < k < \infty \quad \textbf{(Analysis equation)} \tag{5.5}$$

$$x(n) = \sum_{k=\langle N \rangle} c_k e^{j\frac{2\pi kn}{N}} \quad -\infty < n < \infty \quad \textbf{(Synthesis equation)} \tag{5.6}$$

Note: For symmetric $x(n)$, it's convenient to use the following:

- For N odd, we choose $(k$ and $n) = -(N{-}1)/2$ to $(N{-}1)/2$.
- For N even, we choose $(k$ and $n) = -N/2$ to $N/2{-}1$.

Since the coefficients c_k are obtained from the Fourier series expansion in complex form, the resultant spectrum c_k will have two sides. There is an essential feature of Equation (5.5) in which the Fourier series coefficient c_k is periodic of N. We can verify this as follows:

$$c_{k+N} = \frac{1}{N} \sum_{n=0}^{N-1} x(n) e^{-j\frac{2\pi(k+N)n}{N}} = \underbrace{\frac{1}{N} \sum_{n=0}^{N-1} x(n) e^{-j\frac{2\pi kn}{N}}}_{c_k} \underbrace{e^{-j2\pi n}}_{=1} \tag{5.7}$$

Since $e^{-j2\pi n} = \cos(2n\pi) - j\sin(2n\pi) = 1$, it follows that $c_{k+N} = c_k$

Therefore, the two-sided line amplitude spectrum $|c_k|$ is periodic, as shown in Figure 5.2. We note the following points:

a. As displayed in Figure 5.2, only the line spectral portion between frequency $-f_s/2$ and frequency $f_s/2$ (folding frequency) represents the frequency information of the periodic signal.

b. Note that the spectral portion from $f_s/2$ to f_s is a copy of the spectrum in the negative frequency range of $-f_s/2$ to 0 Hz due to the spectrum being periodic for every $N f_0$ Hz. Again, the amplitude spectral components indexed from $f_s/2$ to f_s can be folded at the folding frequency $f_s/2$ to match the magnitude of the spectral components indexed from 0 to $f_s/2$ in terms of $f_s - f$ Hz, where f is in the range of $f_s/2$ to f_s.

Sometimes, we compute the spectrum over the range of 0 to f_s Hz with non-negative indices, that is,

$$c_k = \frac{1}{N} \sum_{n=0}^{N-1} x(n) e^{-j\frac{2\pi kn}{N}} \tag{5.8}$$

$$k = 0,1,2,\ldots,N-1$$

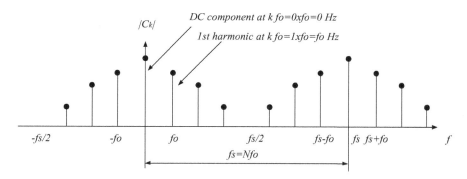

FIGURE 5.2   Periodic discrete signals show the harmonics.

We can apply the same equation to find the negative indexed spectral values if they are required.

c. For the kth harmonic, the frequency is:

$$f = kf_0 \quad Hz \tag{5.9}$$

where f_0, in Hz, is the frequency resolution, which is the frequency of the spacing between the consecutive spectral lines.

5.2   PARSEVAL'S RELATION

For discrete periodic signals, Parseval's relation relates the energy E_g in one period of the sequence to the energy in one period of the Fourier series coefficients as:

$$E_g = \frac{1}{N} \sum_{n=\langle N \rangle} |x(n)|^2 = \sum_{k=\langle N \rangle} |c_k|^2 \tag{5.10}$$

Before presenting several examples illustrating the DTFS, be reminded that the starting values of the indices n and k in the synthesis and analysis equations are *arbitrary* because both $x(n)$ and c_k are N periodic. The range for the indices may thus be chosen to simplify the solution at hand.

Example 5.1

The periodic signal:

$$x(t) = \sin(2\pi t)$$

is sampled using the rate $f_s = 4$ Hz.

    a. Compute the spectrum c_k using the samples in one period.
    b. Plot the two-sided amplitude spectrum $|c_k|$ over the range of –2 to 2 Hz.

SOLUTION

    a. From the analog signal, we can determine the fundamental frequency $\omega_0 = 2\pi$ radians per second and $f_0 = \dfrac{\omega_0}{2\pi} = \dfrac{2\pi}{2\pi} = 1\,\text{Hz}$, and the fundamental period $T_0 = 1$ second. Since we used the sampling interval $T = \dfrac{1}{f_s} = 0.25$ second, we get the sampled signal as:

$$x(t) = x(nT) = \sin(2\pi nT) = \sin(0.5\pi n)$$

and plot the first eight samples, as shown in Figure 5.3.

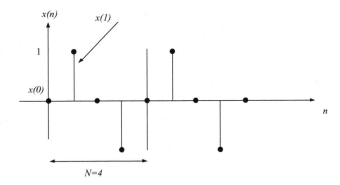

FIGURE 5.3   Input signal $x(n)$.

Choosing the duration of one period, $N = 4$, we get the sample values as follows:

$$x(0) = 0; \; x(1) = 1; \; x(2) = 0; \; x(3) = -1$$

Using the equation of c_k,

$$c_0 = \frac{1}{4}\sum_{n=0}^{3} x(n) = \frac{1}{4}\big(x(0) + x(1) + x(2) + x(3)\big) = \frac{1}{4}(0 + 1 + 0 - 1) = 0$$

$$c_1 = \frac{1}{4}\sum_{n=0}^{3} x(n)\, e^{-j\frac{2\pi 1 n}{4}} = \frac{1}{4}\Big(x(0) + x(1)\, e^{-j\pi/2} + x(2)\, e^{-j\pi} + x(3)\, e^{-j3\pi/2}\Big)$$

$$= \frac{1}{4}\big(x(0) - jx(1) - x(2) + jx(3)\big) = 0 - j(1) - 0 + j(-1) = -j0.5.$$

Similarly, we get:

$$c_2 = \frac{1}{4}\sum_{n=0}^{3} x(n)\, e^{-j\frac{2\pi 2 n}{4}} = 0 \quad \text{and} \quad c_3 = \frac{1}{4}\sum_{n=0}^{3} x(n)\, e^{-j\frac{2\pi 3 n}{4}} = j0.5$$

Using periodicity $c_k = c_{k+N}$, it follows that:

$$c_{-1} = c_3 = j0.5 \quad \text{and} \quad c_{-2} = c_2 = 0$$

b. The amplitude spectrum $|c_k|$ for the digital signal is sketched in Figure 5.4.
   As we know, the spectrum in the range of -2 to 2 Hz presents the information of the sinusoid with a frequency of 1 Hz and a peak value of $2|c_k| = 1$, which is converted from two sides to one side by doubling the spectral value. Note that we do not double the direct-current (DC) component, that is, c_0.

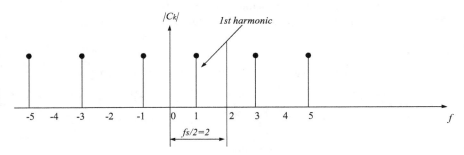

FIGURE 5.4   The amplitude spectrum $|c_k|$.

5.3 DISCREET FOURIER SERIES

For a signal $x[n]$ of finite-length, the DFT is DFS of the periodic extension, $x\sim[n]$, of that signal $x[n]$ and another way to view DFT is that it's a sampling of continuous DTFT.

Given that it is possible to reconstruct the original signal from the sampled signal provided, the sampling is more significant than the Nyquist frequency. We know that the DTFT for the sampled signal is a series of replications of the spectrum of the original signal at frequencies spaced by the sampling frequency. Now, since DTFT is continuous and periodic, we can further breakdown the DTFT into intervals and it will still be possible to reconstruct the DTFT and consequently the original signal. This act of breaking down or sampling the DTFT is called DFT.

Example 5.2

Determine the spectra (DFS) of the signals below:

   a. $x[n] = \cos \sqrt{3}\pi n$
   b. $x[n] = 1, 1, 0, 0$

SOLUTION

   a. The signal is not periodic, and therefore cannot expand into the Fourier series.

   b. $X[k] = \dfrac{1}{N} \displaystyle\sum_{n=0}^{N-1} x(n) e^{-\frac{jk2\pi n}{N}}$

We know that DFS is expressed as

$$X[k] = \frac{1}{N} \sum_{n=0}^{N-1} x(n) e^{-\frac{jk2\pi n}{N}}$$

Here $N = 4$, therefore, we have

$$X[k] = \frac{1}{4} \sum_{n=0}^{3} x(n) e^{-\frac{jk\pi n}{2}}$$

$$X[0] = \frac{1}{4} \sum_{n=0}^{3} x(n) e^{-\frac{jk\pi n}{2}}$$

So, when

$$k = 0 \Rightarrow X(0) = \sum_{n=0}^{3} x(n) \; e^{0}$$

$$= x(0) + x(1) + x(2) + x(3)$$

$$= 1 + 1 + 0 + 0 = 2$$

when

$$k = 1 \Rightarrow X(1) = \sum_{n=0}^{3} x(n) \; e^{-j\pi n/2}$$

$$= x(0) + x(1)e^{-j\pi/2} + x(2)e^{-j\pi} + x(3)e^{-j3\pi/2}$$

$$= 1 + (1)e^{-j\pi/2} + (0)e^{-j\pi} + (0)e^{-j3\pi/2} = 1 - j$$

when

$$k = 2 \Rightarrow X(2) = \sum_{n=0}^{3} x(n) \; e^{-j\pi n}$$

$$= x(0) + x(1)e^{-j\pi} + x(2)e^{-j2\pi} + x(3)e^{-j3\pi}$$

$$= 1 + (1)e^{-j\pi} + (0)e^{-j2\pi} + (0)e^{-j3\pi} = 0$$

when

$$k = 3 \Rightarrow X(3) = \sum_{n=0}^{3} x(n) \; e^{-j3\pi n/2}$$

$$= x(0) + x(1)e^{-j3\pi/2} + x(2)e^{-j3\pi} + x(3)e^{-j9\pi/2}$$

$$= 1 + (1)e^{-j3\pi/2} + (0)e^{-j3\pi} + (0)e^{-j9\pi/2} = 1 + j$$

So, $X(k) = [2, 1 - j, 0, 1 + j]$

Example 5.3

For the periodic signal:

$$x(n) = \left\{ \ldots\ldots, 2, 1, 0, 1, 2, 3, \underset{\uparrow}{2}, 1, 0, 1, 2, \ldots\ldots \right\}.$$

a. Compute the spectrum c_k using the samples from one period.
b. Calculate the energy of the signal using both sides of Parseval's relation.

SOLUTION

a. $c_k = \dfrac{1}{N} \sum_{n=\langle N \rangle} x(n) \, e^{-j\frac{2\pi k n}{N}}$

$c_k = \dfrac{1}{6} \sum_{n=-3}^{2} x(n) \, e^{-j\frac{2\pi k n}{N}} \quad -3 \le k \le 2$

$c_k = \dfrac{1}{6}\left[x(-3)e^{j\frac{6\pi k}{6}} + x(-2)\,e^{j\frac{4\pi k}{6}} + x(-1)\,e^{j\frac{2\pi k}{6}} + x(0) + x(1)\,e^{-j\frac{2\pi k}{6}} + x(2)\,e^{-j\frac{4\pi k}{6}} \right]$

$c_k = \dfrac{1}{6}\left[0\,e^{j\frac{6\pi k}{6}} + 1 e^{j\frac{4\pi k}{6}} + 2\,e^{j\frac{2\pi k}{6}} + 3 + 2\,e^{-j\frac{2\pi k}{6}} + 1 e^{-j\frac{4\pi k}{6}} \right]$

$c_k = \dfrac{1}{6}\left[e^{j\frac{4\pi k}{6}} + 2\,e^{j\frac{2\pi k}{6}} + 3 + 2\,e^{-j\frac{2\pi k}{6}} + e^{-j\frac{4\pi k}{6}} \right]$

Re-arrange the exponential terms:

$$c_k = \dfrac{1}{6}\left[3 + \underbrace{e^{j\frac{4\pi k}{6}} + e^{-j\frac{4\pi k}{6}}}_{2\cos\frac{2\pi k}{3}} + \underbrace{2\,e^{j\frac{2\pi k}{6}} + 2\,e^{-j\frac{2\pi k}{6}}}_{4\cos\frac{\pi k}{3}} \right]$$

$$c_k = \frac{1}{6}\left[3 + 2\cos\frac{2\pi k}{3} + 4\cos\frac{\pi k}{3}\right] \quad -3 \le k \le 2$$

$$c_0 = \frac{9}{6}, \quad c_1 = c_{-1} = \frac{4}{6}, \quad c_2 = c_{-2} = 0, \quad c_{-3} = \frac{1}{6}$$

b. $E_g = \dfrac{1}{N}\displaystyle\sum_{n=\langle N\rangle}|x(n)|^2 = \dfrac{1}{6}\displaystyle\sum_{n=-3}^{2}|x(n)|^2 = \dfrac{1}{6}\left(0^2 + 1^2 + 2^2 + 3^2 + 2^2 + 1^2\right) = \dfrac{19}{6}$

$$E_g = \sum_{k=\langle N\rangle}|c_k|^2 = \sum_{k=-3}^{2}|c_k|^2 = \left[\left(\frac{1}{6}\right)^2 + (0)^2 + \left(\frac{4}{6}\right)^2 + \left(\frac{9}{6}\right)^2 + \left(\frac{4}{6}\right)^2 + (0)^2\right] = \frac{114}{36}$$

$$\therefore \ E_g = \frac{19}{6} = \frac{114}{36}$$

Example 5.4

For the periodic signal:

$$x(n) = \cos\left(\frac{\pi}{8}n + \phi\right)$$

a. Compute the spectrum c_k using the samples from one period.
b. Plot the two-sided amplitude spectrum $|c_k|$.

SOLUTION

a. The fundamental period of $x(t)$ is $N = 16$. Hence, $\Omega_0 = 2\pi/16$.

$$x(n) = \frac{e^{j[(\pi/8)n+\phi]} + e^{-j[(\pi/8)n+\phi]}}{2}$$

$$x(n) = \frac{1}{2}e^{-j\phi}\,e^{-j(\pi/8)n} + \frac{1}{2}e^{j\phi}\,e^{j(\pi/8)n}$$

Compare $x(n)$ with:

$$x(n) = \sum_{k=-8}^{7}c_k\,e^{j\frac{\pi kn}{8}}$$

and by inspection, we get:

$$c_k = \begin{cases} \dfrac{1}{2}e^{-j\phi} & k = -1 \\[2mm] \dfrac{1}{2}e^{j\phi} & k = 1 \\[2mm] 0 & -8 \le k \le 7 \quad \text{and} \quad k \ne \pm 1 \end{cases}.$$

b. The amplitude spectrum $|c_k|$ for the signal is sketched in Figure 5.5.

Example 5.5

Find the DFS coefficients of the periodic sequence with a period of $x(n)$. Plot the magnitudes and phases of the sequence of $x(n)$, within one period, has the form of (Figure 5.6):

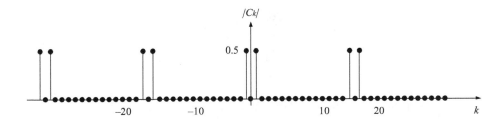

FIGURE 5.5   The amplitude spectrum $|c_k|$.

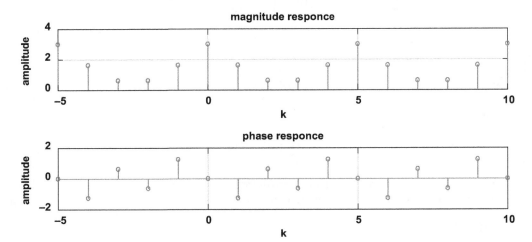

FIGURE 5.6   The magnitude and phase spectrum of Example 5.5.

$$x(n) = \begin{cases} 1 & n = 0,1,2 \\ 0 & n = 3,4 \end{cases}$$

Using DFS $X(k) = e^{-\frac{j2\pi k}{5}}\left[1 + 2\cos\left(\frac{2\pi k}{5}\right)\right]$

ALGORITHM 5.1

```
N=5;
x=[1 1 1 0 0];
k=-N:2*N;
%plot for 3 periods
Xm=abs(1+2.*cos(2*pi.*k/N));
figure;
subplot(2,1,1);
stem(k,Xm,'o')
xlabel('k');
ylabel('amplitude');
title('magnitude response');
%magnitude computation
subplot(2,1,2);
Xa=angle(exp(-2*j*pi.*k/5).*(1+2.*cos(2*pi.*k/N)));
%phase computation
```

```
stem(k,Xa,'o')
xlabel('k');
ylabel('amplitude');
title('phase response');
```

PROBLEMS

5.1   The periodic signal:

$$x(t) = \sin(2\pi t)$$

is sampled using the rate $f_s = 8$ Hz.
a. Compute the spectrum c_k using the samples from one period.
b. Plot the two-sided amplitude spectrum $|c_k|$ over the range of −4 to 4 Hz.

5.2   The periodic signal:

$$x(t) = \cos(\pi t)$$

is sampled using the rate $f_s = 4$ Hz.
a. Compute the spectrum c_k using the samples from one period.
b. Plot the two-sided amplitude spectrum $|c_k|$ over the range of −2 to 2 Hz.

5.3   The periodic signal

$$x(t) = \cos(\pi t)$$

is sampled using the rate $f_s = 6$ Hz.
a. Compute the spectrum c_k using the samples from one period.
b. Plot the two-sided amplitude spectrum $|c_k|$ over the range of −3 to 3 Hz.

5.4   For the periodic signal

$$x(n) = \left\{ \ldots\ldots, 2,3,4,3,2,1,2,3,4,3,2,\ldots\ldots \atop \uparrow \right\}$$

a. Compute the spectrum c_k using the samples from one period.
b. Calculate the energy of the signal using both sides of Parseval's relation.

5.5   For the periodic signal

$$x(n) = \left\{ \ldots\ldots, 1,2,3,3,2,1,1,2,3,3,2,\ldots\ldots \atop \uparrow \right\}$$

a. Compute the spectrum c_k using the samples from one period.
b. Calculate the energy of the signal using both sides of Parseval's relation.

5.6   For the periodic signal

$$x(n) = \cos\left(\frac{\pi}{16}n + \phi\right)$$

a. Compute the spectrum c_k using the samples from one period.
b. Plot the two-sided amplitude spectrum $|c_k|$.

5.7   For the periodic signal

$$x(n) = \sin\left(\frac{\pi}{4}n + \phi\right)$$

   a. Compute the spectrum c_k using the samples from one period.
   b. Plot the two-sided amplitude spectrum $|c_k|$.
5.8 For the periodic signal

$$x(n) = \sin\left(\frac{\pi}{8}n + \phi\right)$$

   c. Compute the spectrum c_k using the samples from one period.
   d. Plot the two-sided amplitude spectrum $|c_k|$.
5.9 For the periodic signal

$$x(n) = \sin\left(\frac{\pi}{16}n + \phi\right)$$

   a. Compute the spectrum c_k using the samples from one period.
   b. Plot the two-sided amplitude spectrum $|c_k|$.
5.10 Determine the DTFS coefficients c_k by inspection for the discrete signal:

$$x(n) = 1 + \sin\left(\frac{1}{12}\pi n + \frac{3\pi}{8}\right)$$

5.11 The periodic digital signal is given by $x(n) = 1 + \sin\left(\frac{\pi n}{8}\right) + 2\cos\left(\frac{\pi n}{4}\right)$

   Find its Fourier series coefficients, and sketch their real and imaginary parts.

5.12 The periodic digital signal is given by $x(n) = 1 + 2\sin\left(\frac{\pi n}{4}\right) + 3\cos\left(\frac{\pi n}{8}\right)$

   Find its Fourier series coefficients, and sketch their real and imaginary parts.

6 Discrete-Time Fourier Transform (DTFT)

The Fourier representation of signals plays an essential part in both continuous and discrete-time signal processing. It delivers a method for mapping signals into another domain in which to operate them. The Fourier representation mainly useful is the property that the convolution operation mapped to multiplication. Also, the Fourier transform provides a different method to understand the systems and signals.

Discrete-Time Fourier Transform (DTFT) is used to analyze the frequency spectrum of a discrete signal with a computer because computers can only handle a finite number of values. This chapter presents analysis a frequency response for a signal using MATLAB, and the Inverse Discrete-Time Fourier Transform(IDTFT), also chapter will submit the Applications of DTFT and analysis of LSI Systems and Solving Difference Equations using DTFT.

6.1 FREQUENCY RESPONSE

Eigen functions of linear shift-invariant systems are sequences that, when input to the system, pass through with only a change in (complex) amplitude. If the input is $x(n)$, the output is $y(n) = \lambda \, x(n)$, where λ, the eigenvalue, generally depends on the input $x(n)$.

Signals of the form

$$x(n) = e^{jn\omega} \quad -\infty < n < \infty \tag{6.1}$$

where ω is a constant, are Eigen functions of LSI systems. Which is shown in the convolution sum:

$$y(n) = h(n) * x(n) = \sum_{k=-\infty}^{\infty} h(k)x(n-k) \tag{6.2}$$

$$= \sum_{k=-\infty}^{\infty} h(k)e^{j\omega(n-k)} = e^{jn\omega} \sum_{k=-\infty}^{\infty} h(k)e^{-jk\omega}$$

$$= H\left(e^{j\omega}\right)e^{jn\omega} \tag{6.3}$$

Thus, the eigenvalue, which we denote as $H\left(e^{j\omega}\right)$, is

$$H\left(e^{j\omega}\right) = \sum_{k=-\infty}^{\infty} h(k)e^{-jk\omega} \tag{6.4}$$

The function $H\left(e^{j\omega}\right)$ is called the *frequency response* of the systems.

Note that, $H\left(e^{j\omega}\right)$ is, in general, a *complex-valued* quantity. Thus, it is written in terms of its *real* and *imaginary* parts.

$$H\left(e^{j\omega}\right) = H_R\left(e^{j\omega}\right) + j\,H_I\left(e^{j\omega}\right) \tag{6.5}$$

or in terms of its *magnitude* and *phase*,

$$H\left(e^{j\omega}\right) = \left|H\left(e^{j\omega}\right)\right| e^{j\phi_h(\omega)} \tag{6.6}$$

and

$$\phi_h(\omega) = \tan^{-1} \frac{H_I\left(e^{j\omega}\right)}{H_R\left(e^{j\omega}\right)} \tag{6.7}$$

A graphical representation that is often used instead of the phase is the group delay, which is defined as follows:

$$\tau_h(\omega) = -\frac{d\phi_h(\omega)}{d\omega} \tag{6.8}$$

In evaluating the group delay, the phase is taken to be a continuous and differentiable function of ω by adding integer multiples of π to the principal value of the phase.

Graphical representations of the frequency response are of great value in the analysis of LSI systems, and plots of the magnitude and phase are commonly used. However, another useful graphical representation is a plot of $20\log\left|H\left(e^{j\omega}\right)\right|$ versus ω. The units on the log magnitude scale are decibels (abbreviated dB). One of the advantages of a *log* magnitude plot is that because the logarithm expands the range for small values of $\left|H\left(e^{j\omega}\right)\right|$, it is useful in displaying the fine detail of the frequency response near zero.

Finally, the inverse DTFT is given by:

$$h(n) = \frac{1}{2\pi} \int_{-\pi}^{\pi} H\left(e^{j\omega}\right) e^{jn\omega} \, d\omega \tag{6.9}$$

Before being given an example, put your mind to the following closed-form expressions for some commonly encountered series:

$$\sum_{n=0}^{N-1} a^n = \frac{1-a^N}{1-a} \tag{6.10}$$

$$\sum_{n=0}^{N-1} na^n = \frac{(N-1)a^{N+1} - Na^N + a}{(1-a)^2} \tag{6.11}$$

$$\sum_{n=0}^{N-1} n = \frac{1}{2} N(N-1) \tag{6.12}$$

$$\sum_{n=0}^{\infty} a^n = \frac{1}{1-a} \quad |a| < 1 \tag{6.13}$$

$$\sum_{n=0}^{\infty} na^n = \frac{a}{(1-a)^2} \quad |a| < 1 \tag{6.14}$$

$$\sum_{n=0}^{N-1} n^2 = \frac{1}{6} N(N-1)(2N-1) \tag{6.15}$$

$$\sum_{k=0}^{n} a^n = \frac{1 - a^{n+1}}{1 - a} \tag{6.16}$$

Example 6.1

Consider the LSI system with unit sample response

$$h(n) = a^n u(n)$$

where α is a real number with $|a| < 1$. The frequency response is

$$H\left(e^{j\omega}\right) = \sum_{n=-\infty}^{\infty} h(n) e^{-jn\omega} = \sum_{n=0}^{\infty} a^n e^{-jn\omega}$$

$$= \sum_{n=0}^{\infty} \left(a e^{-j\omega}\right)^n = \frac{1}{1 - a e^{-j\omega}}$$

The squared magnitude of the frequency response is

$$\left|H\left(e^{j\omega}\right)\right|^2 = H\left(e^{j\omega}\right) H^*\left(e^{j\omega}\right)$$

$$= \frac{1}{1 - a e^{-j\omega}} \cdot \frac{1}{1 - a e^{j\omega}} = \frac{1}{1 + a^2 - 2a\cos\omega}$$

and the phase is

$$\phi_h(\omega) = \tan^{-1} \frac{H_I\left(e^{j\omega}\right)}{H_R\left(e^{j\omega}\right)} = \tan^{-1} \frac{-a\sin\omega}{1 - a\cos\omega}$$

Finally, the group delay is found by differentiating the phase. The result is

$$\tau_h(\omega) = \frac{a^2 - a\cos\omega}{1 + a^2 - 2a\,\cos\omega}$$

Example 6.2

For a system with a frequency response given by

$$H\left(e^{j\omega}\right) = \begin{cases} 1 & |\omega| \le \omega_c \\ 0 & \omega_c < |\omega| \le \pi \end{cases}$$

(this system is referred to as an ideal low-pass filter), the unit sample response is

$$h(n) = \frac{1}{2\pi} \int_{-\omega_c}^{\omega_c} e^{jn\omega} \, d\omega = \frac{1}{2j\pi n} \left[e^{jn\omega_c} - e^{-jn\omega_c} \right] = \frac{\sin n\omega_c}{\pi n}$$

Note that this system is noncausal (it is also unstable) and, therefore, unrealizable.

6.2   DTFT FOR ANY DISCRETE SIGNAL

The frequency response of a linear shift-invariant system is found by multiplying $h(n)$ by a complex exponential, $e^{-jn\omega}$, and summing over n. The Discrete-Time Fourier transform of a sequence, $x(n)$, is defined in the same way,

$$X\left(e^{j\omega}\right) = \sum_{n=-\infty}^{\infty} x(n)e^{-jn\omega} \tag{6.17}$$

Thus, the frequency response of a linear shift-invariant system, $H\left(e^{j\omega}\right)$ is the DTFT of the unit sample response, $h(n)$. For the DTFT of a sequence to exist, the summation in $X\left(e^{j\omega}\right)$ must converge. So, in turn, this requires that $x(n)$ be summable:

$$\sum_{n=-\infty}^{\infty} |x(n)| = S < \infty \tag{6.18}$$

Example 6.3

Find the DTFT of the sequence

$$x(n) = -a^n u(-n-1) \quad |a| > 1$$

SOLUTION

$$X\left(e^{j\omega}\right) = \sum_{n=-\infty}^{\infty} x_2(n)e^{-jn\omega} = -\sum_{n=-\infty}^{-1} a^n e^{-jn\omega}$$

Changing the limits on the sum, we have

$$X\left(e^{j\omega}\right) = -\sum_{n=1}^{\infty} a^{-n} e^{jn\omega} = -\sum_{n=0}^{\infty} \left(a^{-1} e^{-j\omega}\right)^n + 1$$

If $|a| > 1$, this sum is (Table 6.1)

$$X\left(e^{j\omega}\right) = \frac{1}{1-a^{-1}e^{j\omega}} + 1$$

$$X\left(e^{j\omega}\right) = \frac{1}{1-ae^{-j\omega}} \quad \text{for } |a| > 1$$

6.3   INVERSE DTFT

Given $X\left(e^{j\omega}\right)$, the sequence $x(n)$ is recovered using the inverse DTFT,

$$x(n) = \frac{1}{2\pi} \int_{-\pi}^{\pi} X\left(e^{j\omega}\right) e^{jn\omega} d\omega \tag{6.19}$$

The inverse DTFT is viewed as a decomposition of $x(n)$ into a linear combination of all complex exponentials. Table 6.1 contains a list of some useful DTFT pairs.

Note that in some cases $x(n)$ is not summable. But by allowing the DTFT to contain impulses, we may consider the DTFT of sequences that contain complex exponentials.

Example 6.4

Find the inverse DTFT of:

$$X\left(e^{j\omega}\right) = \delta(\omega - \omega_0)$$

TABLE 6.1
DTFT Sequences

Sequence	DTFT
$a\delta(n)$	a
$a\,\delta(n-n_o)$	$a\,e^{-jn_o\omega}$
a	$a\,2\pi\,\delta(\omega)$
$a\,e^{jn\omega_o}$	$a\,2\pi\,\delta(\omega-\omega_o)$
$a^n\,u(n), |a|<1$	$\dfrac{1}{1-ae^{-j\omega}}$
$-a^n\,u(-n-1), |a|>1$	$\dfrac{1}{1-ae^{-j\omega}}$
$(n+1)a^n\,u(n), |a|<1$	$\dfrac{1}{\left(1-ae^{-j\omega}\right)^2}$
$\cos n\omega_o$	$a\,\pi\,\delta(\omega+\omega_o)+a\,\pi\,\delta(\omega-\omega_o)$

We have

$$x(n) = \frac{1}{2\pi}\int_{-\pi}^{\pi} X\left(e^{j\omega}\right)e^{jn\omega}\,d\omega = \frac{1}{2\pi}e^{jn\omega_0}$$

Example 6.5

Find the inverse DTFT of:

$$X\left(e^{j\omega}\right) = \pi\,\delta(\omega-2)+\pi\,\delta(\omega+2)$$

SOLUTION

Computing the inverse DTFT, we find

$$x(n) = \frac{1}{2}e^{j2n}+\frac{1}{2}e^{-j2n} = \cos(2n)$$

6.4   INTERCONNECTION OF SYSTEMS

1. Cascade systems
   A cascade of two linear shift-invariant systems is shown in Figure 6.1.

$$h(n) = h_1(n)*h_2(n) \tag{6.20}$$

and a frequency response

$$H\left(e^{j\omega}\right) = H_1\left(e^{j\omega}\right)H_2\left(e^{j\omega}\right) \tag{6.21}$$

The log magnitude is the *sum* of the log magnitudes of the individual systems,

$$20\log\left|H\left(e^{j\omega}\right)\right| = 20\log\left|H_1\left(e^{j\omega}\right)\right| + 20\log\left|H_2\left(e^{j\omega}\right)\right| \tag{6.22}$$

FIGURE 6.1   Cascade systems.

and the phase and group delay are additive,

$$\phi(\omega) = \phi_1(\omega) + \phi_2(\omega) \tag{6.23}$$

$$\tau(\omega) = \tau_1(\omega) + \tau_2(\omega) \tag{6.24}$$

2. Parallel systems: A parallel connection between two linear shift-invariant systems. As shown in Figure 6.2

$$h(n) = h_1(n) + h_2(n) \tag{6.25}$$

The frequency response of the parallel network is

$$H\left(e^{j\omega}\right) = H_1\left(e^{j\omega}\right) + H_2\left(e^{j\omega}\right) \tag{6.26}$$

3. A feedback network is commonly found in control applications. Figure 6.3 shows a feedback network.
   This network may be analyzed as follows. With

$$g(n) = x(n) + b(n) * y(n) \tag{6.27}$$

$$y(n) = a(n) * g(n) \tag{6.28}$$

The frequency response of this system, if it exists, is

$$H\left(e^{j\omega}\right) = \frac{Y\left(e^{j\omega}\right)}{X\left(e^{j\omega}\right)} = \frac{A\left(e^{j\omega}\right)}{1 - A\left(e^{j\omega}\right)B\left(e^{j\omega}\right)} \tag{6.29}$$

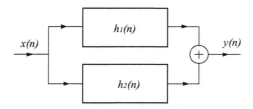

FIGURE 6.2   Parallel systems.

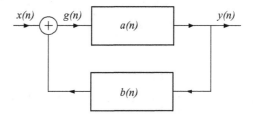

FIGURE 6.3   Feedback network.

6.5   DTFT PROPERTIES

There are several features of the DTFT that may be used to simplify the evaluation of the DTFT and its inverse. A summary of the DTFT properties appears in Table 6.2.

6.6   APPLICATIONS OF DTFT

In this section, we present some applications of the DTFT in the discrete-time signal analysis. These include finding the frequency response of an LSI system that is described by a difference equation, performing convolutions, solving difference equations that have zero initial conditions, and designing inverse systems.

6.7   LSI SYSTEMS AND DIFFERENCE EQUATIONS

A significant part of LSI systems contains those whose input, $x(n)$, and output, $y(n)$, are related by a linear constant-coefficient difference equation.

Example 6.5

Consider the linear shift-invariant system characterized by the second-order linear constant-coefficient difference equation:

$$y(n) = 1.3433\,y(n-1) - 0.9025\,y(n-2) + x(n) - 1.4142x(n-1) + x(n-2)$$

The frequency response can be found through inspection without solving the difference equation for $h(n)$ as follows:

$$H\left(e^{j\omega}\right) = \frac{1 - 1.4142e^{-j\omega} + e^{-2j\omega}}{1 - 1.3433e^{-j\omega} + 0.9025e^{-2j\omega}}$$

TABLE 6.2
DTFT Properties

Property	Sequence	DTFT
Linearity	$a_1 x(n) + a_2 y(n)$	$a_1 X\left(e^{j\omega}\right) + a_2 Y\left(e^{j\omega}\right)$
Convolution	$a_1 x(n) * a_2 y(n)$	$a_1 a_2\, X\left(e^{j\omega}\right) \cdot Y\left(e^{j\omega}\right)$
Time-reversal	$a\, x(-n)$	$a X\left(e^{-j\omega}\right)$
Shift	$a\, x(n - n_o)$	$a e^{-jn_o\omega}\, X\left(e^{j\omega}\right)$
Modulation	$a e^{jn\omega_o}\, x(n)$	$a X\left(e^{j(\omega-\omega_o)}\right)$
Conjugation	$a\, x(n)^*$	$a X^*\left(e^{-j\omega}\right)$
Derivative	$a\, n\, x(n)$	$j\dfrac{dX\left(e^{j\omega}\right)}{d\omega}$
Multiplication	$a_1 x(n)\, a_2 y(n)$	$a_1 a_2\, \dfrac{1}{2\pi}\displaystyle\int_{-\pi}^{\pi} X\left(e^{j\omega}\right) \cdot Y\left(e^{j(\omega-\theta)}\right) d\theta$

Performing Convolutions

The DTFT provides another method for performing convolutions in the time domain, because the DTFT maps convolution in the time domain into multiplication in the frequency domain,

Example 6.6

If the unit sample response of an LSI system is:

$$h(n) = a^n u(n)$$

Let us find the response of the system to the input $x(n) = \beta^n u(n)$ where $|a| < 1$, $|\beta| < 1$, and $a \neq \beta$. Because the output of the system is the convolution of $x(n)$ with $h(n)$

$$y(n) = h(n) * x(n)$$

And the DTFT of $y(n)$ is

$$Y\left(e^{j\omega}\right) = H\left(e^{j\omega}\right) X\left(e^{j\omega}\right) = \frac{1}{1 - ae^{-j\omega}} \cdot \frac{1}{1 - \beta e^{-j\omega}}$$

Therefore, all that is required is to find the inverse DTFT of $Y\left(e^{j\omega}\right)$. This may be done easily by expanding $Y\left(e^{j\omega}\right)$ as follows:

$$Y\left(e^{j\omega}\right) = \frac{1}{\left(1 - ae^{-j\omega}\right)\left(1 - \beta e^{-j\omega}\right)} = \frac{A}{1 - ae^{-j\omega}} + \frac{B}{1 - \beta e^{-j\omega}}$$

where A and B are constants that are to be determined. Expressing the right-hand side of this expansion over a common denominator,

$$\frac{1}{\left(1 - ae^{-j\omega}\right)\left(1 - \beta e^{-j\omega}\right)} = \frac{(A+B) - (A\beta + Ba)e^{-j\omega}}{\left(1 - ae^{-j\omega}\right)\left(1 - \beta e^{-j\omega}\right)}$$

and equating the coefficients, the constants A and B may be found by solving the pair of equations:

$$A + B = 1$$

$$A\beta + Ba = 0$$

The result is

$$A = \frac{a}{a - \beta}$$

$$B = -\frac{\beta}{a - \beta}$$

Therefore,

$$Y\left(e^{j\omega}\right) = \frac{\dfrac{a}{a - \beta}}{1 - ae^{-j\omega}} - \frac{\dfrac{a}{a - \beta}}{1 - \beta e^{-j\omega}}$$

and it follows that the inverse DTFT is

$$y(n) = \left[\frac{a}{a - \beta} a^n - \frac{\beta}{a - \beta} \beta^n \right] u(n)$$

6.8   SOLVING DIFFERENCE EQUATIONS USING DTFT

In the previous chapter, we looked at methods for solving difference equations in the "time domain." The DTFT may be used to solve difference equations in the "frequency domain" provided that the initial conditions are zero. The procedure is to transform the difference equation into the frequency domain by taking the DTFT of each term in the equation, solving it for the desired term, and finding the inverse DTFT.

Example 6.7

Solve the below for $y(n)$ assuming zero initial conditions,

$$y(n) - 0.25\, y(n-1) = x(n) - x(n-2) \quad \text{for } x(n) = \delta(n).$$

We begin by taking the DTFT of each term in the difference equation:

$$Y\left(e^{j\omega}\right) - 0.25 e^{-j\omega} Y\left(e^{j\omega}\right) = X\left(e^{j\omega}\right) - e^{-2j\omega} X\left(e^{j\omega}\right)$$

Because of the DTFT of $x(n)$ is $X\left(e^{j\omega}\right) = 1$,

$$Y\left(e^{j\omega}\right) = \frac{1 - e^{-2j\omega}}{1 - 0.25 e^{-j\omega}} = \frac{1}{1 - 0.25 e^{-j\omega}} - \frac{e^{-2j\omega}}{1 - 0.25 e^{-j\omega}}$$

Using the DTFT pair

$$(0.25)^n\, u(n) \overset{\text{DTFT}}{\Longleftrightarrow} \frac{1}{1 - 0.25 e^{-j\omega}}$$

the inverse DTFT of $Y\left(e^{j\omega}\right)$ may be easily found using the linearity and shift properties,

$$y(n) = (0.25)^n\, u(n) - (0.25)^{n-2}\, u(n-2)$$

Example 6.8

Find the frequency response for the following LSI system

a) $h(n) = \delta(n) + 2\delta(n-1) + 3\delta(n-3)$

b) $h(n) = \left(\dfrac{1}{4}\right)^{n+3} u(n-3)$

SOLUTION

$$H\left(e^{-j\omega}\right) = \sum_{n=-\infty}^{\infty} h(n) e^{-jn\omega}$$

$$H\left(e^{-j\omega}\right) = \sum_{n=-\infty}^{\infty} \left[\delta(n) + 2\delta(n-1) + 3\delta(n-3)\right] e^{-jn\omega}$$

Since,

$$\sum_{n=-\infty}^{\infty} \left[\delta(n-n_0)\right] e^{-jn\omega} = e^{-jn_0\omega}$$

So,

$$H\left(e^{-j\omega}\right) = 1 + 2e^{-j\omega} + 3e^{-j3\omega}$$

$$H\left(e^{-j\omega}\right) = \sum_{n=-\infty}^{\infty} h(n)e^{-jn\omega}$$

$$H\left(e^{-j\omega}\right) = \sum_{n=0}^{\infty}\left[\left(\frac{1}{4}\right)^{n+3} u(n-3)\right]e^{-jn\omega}$$

$$= \sum_{n=0}^{\infty}\left[\left(\frac{1}{4}\right)^{n+3}\right]e^{-j(n+3)\omega}$$

$$= \left(\frac{1}{4}\right)^{3} e^{-j3\omega} \sum_{n=0}^{\infty}\left[\left(\frac{1}{4}\right)^{n}\right]e^{-jn\omega}$$

$$= \left(\frac{1}{4}\right)^{3} e^{-j3\omega} \sum_{n=0}^{\infty}\left(\frac{1}{4}e^{-j\omega}\right)^{n}$$

Using the geometric series, we find

$$= \left(\frac{1}{4}\right)^{3} e^{-j3\omega} \frac{1}{1-\frac{1}{4}e^{-j\omega}}$$

Example 6.9

Find the frequency response of an LSI system described by the difference equation

$$y(n) - 0.75\,y(n-1) = x(n) + 2x(n-1) + 3x(n-2)$$

SOLUTION

To find the frequency, we find DTFT for the difference equation

$$\left(1 - 0.75e^{-j\omega}\right)Y\left(e^{j\omega}\right) = \left(1 + 2e^{-j\omega} + 3e^{-j2\omega}\right)X\left(e^{j\omega}\right)$$

$$H\left(e^{j\omega}\right) = \frac{Y\left(e^{j\omega}\right)}{X\left(e^{j\omega}\right)} = \frac{1 + 2e^{-j\omega} + 3e^{-j2\omega}}{1 - 0.75e^{-j\omega}}$$

Example 6.10

Find the DTFT of the following sequence

a) $x_1 = (0.8)^{n}\,u(n+2)$

$$X_1\left(e^{-j\omega}\right) = \sum_{n=-\infty}^{\infty} x_1(n)e^{-jn\omega}$$

$$= \sum_{n=-\infty}^{\infty}\left[(0.8)^{n}\,u(n+2)\right]e^{-jn\omega}$$

$$= \sum_{n=-2}^{\infty} \left[(0.8)^n \, e^{-jn\omega} \right]$$

$$= \sum_{n=-2}^{\infty} \left(0.8 \, e^{-j\omega} \right)^n$$

$$= \left(0.8 \, e^{-jn\omega} \right)^{-2} \sum_{n=0}^{\infty} \left(0.8 \, e^{-j\omega} \right)^n$$

$$= \frac{25}{16} e^{j2\omega} \sum_{n=0}^{\infty} \left(0.8 \, e^{-j\omega} \right)^n$$

$$= \frac{25}{16} e^{j2\omega} \frac{1}{1 - 0.8 \, e^{-j\omega}}$$

b) $x_2(n) = a^n \sin(2n) u(n)$

$$X_2\left(e^{-j\omega}\right) = \sum_{n=-\infty}^{\infty} x_2(n) e^{-jn\omega}$$

$$X_2\left(e^{-j\omega}\right) = \sum_{n=-\infty}^{\infty} \left[a^n \sin(2n) u(n) \right] e^{-jn\omega}$$

$$= \sum_{n=0}^{\infty} \left[a^n \sin(2n) \right] e^{-jn\omega}$$

$$= \sum_{n=0}^{\infty} a^n \left[\frac{e^{j2n} - e^{-j2n}}{2j} \right] e^{-jn\omega}$$

$$= \frac{1}{2j} \sum_{n=0}^{\infty} a^n \left(e^{j2n} - e^{-j2n} \right) e^{-jn\omega}$$

$$= \frac{1}{2j} \sum_{n=0}^{\infty} \left(a^n \, e^{j2n} \, e^{-jn\omega} - a^n e^{-j2n} \, e^{-jn\omega} \right)$$

$$= \frac{1}{2j} \sum_{n=0}^{\infty} \left(a^n \, e^{j2n} \, e^{-jn\omega} \right) - \frac{1}{2j} \sum_{n=0}^{\infty} \left(a^n e^{-j2n} \, e^{-jn\omega} \right)$$

$$= \frac{1}{2j} \sum_{n=0}^{\infty} (a e^{-j(\omega-2)})^n - \frac{1}{2j} \sum_{n=0}^{\infty} \left(a e^{-j(\omega+2)} \right)^n$$

$$= \frac{1}{2j} \left[\frac{1}{1 - a e^{-j(\omega-2)}} - \frac{1}{1 - a e^{-j(\omega+2)}} \right]$$

$$= \frac{1}{2j} \left[\frac{a \sin 2 \, e^{-j\omega}}{1 - 2a \cos 2 \, e^{-j\omega} + a^2 e^{-j2\omega}} \right]$$

c) $x_3(n) = \begin{cases} (0.25)^n & \text{for } n \geq 0 \text{ and even only} \\ 0 & \text{otherwise} \end{cases}$

$$X_3\left(e^{-j\omega}\right) = \sum_{n=-\infty}^{\infty} x_3(n) e^{-jn\omega}$$

$$X_3\left(e^{-j\omega}\right) = \sum_{n=0,2,4,\dots}^{\infty} (0.25)^n\, e^{-jn\omega}$$

$$X_3\left(e^{-j\omega}\right) = \sum_{n=0}^{\infty} (0.25)^{2n}\, e^{-j2n\omega}$$

$$= \sum_{n=0}^{\infty} \left(\frac{1}{16} e^{-j2\omega}\right)^n$$

$$= \frac{1}{1 - \dfrac{1}{16} e^{-j2\omega}}$$

Example 6.11

Find the DTFT of the following sequence (Figure 6.4).

 a) Find the transfer function $H\left(e^{j\omega}\right)$
 b) Find the frequency response if

$$h_1(n) = \delta(n) + \delta(n-1) + \delta(n-2) = h_2(n)$$

$$h_3(n) = (0.5)^n\, u(n) = h_4(n)$$

SOLUTION

 a) $G_1\left(e^{j\omega}\right) = H_1\left(e^{j\omega}\right) \cdot H_2\left(e^{j\omega}\right)$

$$G_2\left(e^{j\omega}\right) = H_3\left(e^{j\omega}\right) \cdot H_4\left(e^{j\omega}\right)$$

$$H\left(e^{j\omega}\right) = \frac{Y\left(e^{j\omega}\right)}{X\left(e^{j\omega}\right)} = G_1\left(e^{j\omega}\right) + G_2\left(e^{j\omega}\right)$$

$$H\left(e^{j\omega}\right) = \frac{Y\left(e^{j\omega}\right)}{X\left(e^{j\omega}\right)} = H_1\left(e^{j\omega}\right) \cdot H_2\left(e^{j\omega}\right) + H_3\left(e^{j\omega}\right) \cdot H_4\left(e^{j\omega}\right)$$

 b) $H_1\left(e^{j\omega}\right) = 1 + e^{-j\omega} + e^{-j2\omega} = H_2\left(e^{j\omega}\right)$

So,

$$G_1\left(e^{j\omega}\right) = \left(1 + e^{-j\omega} + e^{-j2\omega}\right)^2$$

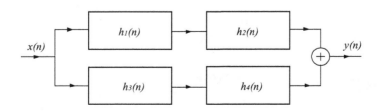

FIGURE 6.4 System of Example 6.11.

$$H_3\left(e^{j\omega}\right)=\frac{1}{1-0.5e^{-j\omega}}=H_4\left(e^{j\omega}\right)$$

So,

$$G_2\left(e^{j\omega}\right)=\left(\frac{1}{1-0.5e^{-j\omega}}\right)^2$$

$$H\left(e^{j\omega}\right)=\frac{Y\left(e^{j\omega}\right)}{X\left(e^{j\omega}\right)}=H_1\left(e^{j\omega}\right)\cdot H_2\left(e^{j\omega}\right)+H_3\left(e^{j\omega}\right)\cdot H_4\left(e^{j\omega}\right)$$

Therefore,

$$H\left(e^{j\omega}\right)=\frac{Y\left(e^{j\omega}\right)}{X\left(e^{j\omega}\right)}=\left(1+e^{-j\omega}+e^{-j2\omega}\right)^2+\left(\frac{1}{1-0.5e^{-j\omega}}\right)^2$$

Example 6.12

Find the DTFT of the following sequence (Figure 6.5).

a) Find the transfer function $H\left(e^{j\omega}\right)$
b) Find the frequency response if

$$h_1(n)=\delta(n)+2\delta(n-2)+3\delta(n-2)=h_2(n)$$

$$h_3(n)=(0.75)^n\,u(n)=h_4(n)$$

SOLUTION

a) $G_1\left(e^{j\omega}\right)=H_2\left(e^{j\omega}\right)\cdot H_3\left(e^{j\omega}\right)$

$$G_2\left(e^{j\omega}\right)=G_1\left(e^{j\omega}\right)+H_4\left(e^{j\omega}\right)$$

$$G_2\left(e^{j\omega}\right)=H_2\left(e^{j\omega}\right)\cdot H_3\left(e^{j\omega}\right)+H_4\left(e^{j\omega}\right)$$

$$H\left(e^{j\omega}\right)=\frac{Y\left(e^{j\omega}\right)}{X\left(e^{j\omega}\right)}=H_1\left(e^{j\omega}\right)\cdot G_2\left(e^{j\omega}\right)$$

$$H\left(e^{j\omega}\right)=\frac{Y\left(e^{j\omega}\right)}{X\left(e^{j\omega}\right)}=H_1\left(e^{j\omega}\right)\cdot\left[H_2\left(e^{j\omega}\right)\cdot H_3\left(e^{j\omega}\right)+H_4\left(e^{j\omega}\right)\right]$$

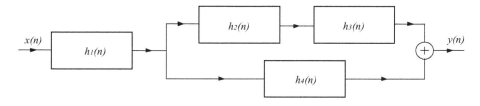

FIGURE 6.5 System of Example 6.12.

$$H\left(e^{j\omega}\right) = \frac{Y\left(e^{j\omega}\right)}{X\left(e^{j\omega}\right)} = H_1\left(e^{j\omega}\right) \cdot H_2\left(e^{j\omega}\right) \cdot H_3\left(e^{j\omega}\right) + H_1\left(e^{j\omega}\right) \cdot H_4\left(e^{j\omega}\right)$$

b) $H_1\left(e^{j\omega}\right) = 1 + 2e^{-j\omega} + 3e^{-j2\omega} = H_2\left(e^{j\omega}\right)$

and

$$H_3\left(e^{j\omega}\right) = \frac{1}{1 - 0.75\,e^{-j\omega}} = H_4\left(e^{j\omega}\right)$$

$$H\left(e^{j\omega}\right) = \frac{Y\left(e^{j\omega}\right)}{X\left(e^{j\omega}\right)} = H_1\left(e^{j\omega}\right) \cdot H_2\left(e^{j\omega}\right) \cdot H_3\left(e^{j\omega}\right) + H_1\left(e^{j\omega}\right) \cdot H_4\left(e^{j\omega}\right)$$

Therefore,

$$H\left(e^{j\omega}\right) = \frac{Y\left(e^{j\omega}\right)}{X\left(e^{j\omega}\right)} = \left(1 + 2e^{-j\omega} + 3e^{-j2\omega}\right)^2 \cdot \frac{1}{1 - 0.75\,e^{-j\omega}}$$

$$+ \left(1 + 2e^{-j\omega} + 3e^{-j2\omega}\right) \cdot \frac{1}{1 - 0.75\,e^{-j\omega}}$$

$$H\left(e^{j\omega}\right) = \frac{Y\left(e^{j\omega}\right)}{X\left(e^{j\omega}\right)} = \frac{\left(1 + 2e^{-j\omega} + 3e^{-j2\omega}\right)}{1 - 0.75\,e^{-j\omega}} \cdot \left[2 + 2e^{-j\omega} + 3e^{-j2\omega}\right]$$

Example 6.13

A LSI system is described by a difference equation,

$$y(n) = 0.1y(n-1) + a\,x(n)$$

a) Find the gain value α so that the magnitude of the transfer function is equal to that value at the dc condition.
b) Find the half power point.

SOLUTION

$$y(n) = 0.1y(n-1) + a\,x(n)$$

Take DTFT

$$Y\left(e^{j\omega}\right) = 0.1e^{-j\omega}Y\left(e^{j\omega}\right) + aX\left(e^{j\omega}\right)$$

$$\left(1 - 0.1e^{-j\omega}\right)Y\left(e^{j\omega}\right) = aX\left(e^{j\omega}\right)$$

$$H\left(e^{j\omega}\right) = \frac{Y\left(e^{j\omega}\right)}{X\left(e^{j\omega}\right)} = \frac{a}{\left(1 - 0.1e^{-j\omega}\right)}$$

$$H\left(e^{j\omega}\right) = \frac{Y\left(e^{j\omega}\right)}{X\left(e^{j\omega}\right)} = \frac{a}{\left(1 - 0.1\left(\cos\omega - j\sin\omega\right)\right)}$$

$$H\left(e^{j\omega}\right) = \frac{Y\left(e^{j\omega}\right)}{X\left(e^{j\omega}\right)} = \frac{a}{\left(1 - 0.1\cos\omega\right) - j\,0.1\sin\omega}$$

$$\left|H\left(e^{j\omega}\right)\right| = \frac{a}{\sqrt{\left(1-0.1\cos\omega\right)^2 + \left(0.1\sin\omega\right)^2}}$$

$$\left|H\left(e^{j\omega}\right)\right| = \frac{a}{\sqrt{1.01 - 0.2\cos\omega}}$$

At dc $\omega = 0$ so,

$$1 = \frac{a}{\sqrt{1.01 - 0.2}}$$

$$a = \mathbf{0.9}$$

Therefore,

$$H\left(e^{j\omega}\right) = \frac{Y\left(e^{j\omega}\right)}{X\left(e^{j\omega}\right)} = \frac{0.9}{\left(1 - 0.1e^{-j\omega}\right)}$$

Example 6.14

A LTI system is described by a difference equation

$$y(n) = 0.8\,y(n-1) + a\,x(n)$$

a) Find the gain value α so that the magnitude of the transfer function is equal to that value at the dc condition.
b) Find the half power point.

SOLUTION

$$y(n) = 0.8\,y(n-1) + a\,x(n)$$

Take DTFT

$$Y\left(e^{j\omega}\right) = 0.8\,e^{-j\omega}Y\left(e^{j\omega}\right) + a\,X\left(e^{j\omega}\right)$$

$$\left(1 - 0.8\,e^{-j\omega}\right)Y\left(e^{j\omega}\right) = a\,X\left(e^{j\omega}\right)$$

$$H\left(e^{j\omega}\right) = \frac{Y\left(e^{j\omega}\right)}{X\left(e^{j\omega}\right)} = \frac{a}{\left(1 - 0.8\,e^{-j\omega}\right)}$$

$$H\left(e^{j\omega}\right) = \frac{Y\left(e^{j\omega}\right)}{X\left(e^{j\omega}\right)} = \frac{a}{\left(1 - 0.8\left(\cos\omega - j\sin\omega\right)\right)}$$

$$H\left(e^{j\omega}\right) = \frac{Y\left(e^{j\omega}\right)}{X\left(e^{j\omega}\right)} = \frac{a}{\left(1 - 0.8\cos\omega\right) - j\,0.8\sin\omega}$$

$$\left|H\left(e^{j\omega}\right)\right| = \frac{a}{\sqrt{\left(1 - 0.8\cos\omega\right)^2 + \left(0.8\sin\omega\right)^2}}$$

$$\left|H\left(e^{j\omega}\right)\right| = \frac{a}{\sqrt{1.64 - 1.6\cos\omega}}$$

At dc $\omega = 0$ so,

$$1 = \frac{a}{\sqrt{1.64 - 1.6}}$$

$$a = \mathbf{0.2}$$

Therefore,

$$H\left(e^{j\omega}\right) = \frac{Y\left(e^{j\omega}\right)}{X\left(e^{j\omega}\right)} = \frac{0.2}{\left(1 - 0.8e^{-j\omega}\right)}$$

$$\frac{1}{2} = \frac{0.2}{\sqrt{1.64 - 1.6\cos\omega}}$$

Solving this, we get $\omega = 0.124\,\pi\mathrm{rad}$.

Example 6.15

Write a difference equation to implement a system with a frequency response

$$H\left(e^{j\omega}\right) = \frac{1 - e^{-j\omega} + 0.8e^{-j2\omega}}{1 + 0.7e^{-j\omega} - 0.9e^{-j2\omega}}$$

SOLUTION

$$H\left(e^{j\omega}\right) = \frac{Y\left(e^{j\omega}\right)}{X\left(e^{j\omega}\right)} = \frac{1 - e^{-j\omega} + 0.8e^{-j2\omega}}{1 + 0.7e^{-j\omega} - 0.9e^{-j2\omega}}$$

After cross-multiplying, we have

$$\left[1 + 0.7e^{-j\omega} - 0.9e^{-j2\omega}\right]Y\left(e^{j\omega}\right) = \left[1 - e^{-j\omega} + 0.8e^{-j2\omega}\right]X\left(e^{j\omega}\right)$$

Taking the inverse DTFT of each term gives the desired difference equation

$$y(n) + 0.7y(n-1) - 0.9y(n-2) = x(n) - x(n-1) + 0.8x(n-2)$$

$$y(n) = -0.7y(n-1) + 0.9y(n-2) + x(n) - x(n-1) + 0.8x(n-2)$$

Example 6.16

Find the group delay for each of the following systems, where α is a real number.

a) For the sequence, the frequency response

$$H\left(e^{jw}\right) = 1 - 2e^{-jw}$$

$$H\left(e^{j\omega}\right) = 1 - 2e^{-j\omega}$$

$$H\left(e^{j\omega}\right) = 1 - 2\left(\cos\omega - j\sin\omega\right)$$

Therefore, the phase is

$$\varphi(\omega) = \tan^{-1}\frac{2\sin\omega}{1 - 2\cos\omega}$$

$$\tau = \frac{d}{d\omega}\varphi(\omega) = \frac{d}{d\omega}\tan^{-1}\frac{2\sin\omega}{1-2\cos\omega}$$

$$\tau = \frac{d}{d\omega}\varphi(\omega) = \frac{1}{1+\left(\dfrac{2\sin\omega}{1-2\cos\omega}\right)^2}\frac{d}{d\omega}\left(\frac{2\sin\omega}{1-2\cos\omega}\right)$$

Therefore,

$$\tau = \frac{1}{1+\left(\dfrac{2\sin\omega}{1-2\cos\omega}\right)^2}\left(\frac{(1-2\cos\omega)2\cos\omega-(2\sin\omega)^2}{(1-2\cos\omega)^2}\right)$$

So, after simplification, becomes

$$\tau = \frac{2(2-\cos\omega)}{(5-4\cos\omega)^2}$$

Example 6.17

Find the group delay for each of the following systems, where α is a real number.

a) For the sequence, the frequency response is

$$H\left(e^{j\omega}\right) = 1-\beta e^{-j\omega}$$

$$H\left(e^{j\omega}\right) = 1-\beta e^{-j\omega}$$

$$H\left(e^{j\omega}\right) = 1-\beta(\cos\omega - j\sin\omega)$$

Therefore, the phase is

$$\phi(\omega) = \tan^{-1}\frac{\beta\sin\omega}{1-\beta\cos\omega}$$

$$\tau = \frac{d}{d\omega}\phi(\omega) = \frac{d}{d\omega}\tan^{-1}\frac{\beta\sin\omega}{1-\beta\cos\omega}$$

$$\tau = \frac{d}{d\omega}\phi(\omega) = \frac{1}{1+\left(\dfrac{\beta\sin\omega}{1-\beta\cos\omega}\right)^2}\frac{d}{d\omega}\left(\frac{\beta\sin\omega}{1-\beta\cos\omega}\right)$$

Therefore,

$$\tau = \frac{1}{1+\left(\dfrac{\beta\sin\omega}{1-\beta\cos\omega}\right)^2}\left(\frac{(1-\beta\cos\omega)\beta\cos\omega-(\beta\sin\omega)^2}{(1-\beta\cos\omega)^2}\right)$$

So, after simplification, it becomes

$$\tau = \frac{\beta(\beta-\cos\omega)}{(1+\beta^2-2\beta\cos\omega)^2}$$

Example 6.18

$$\left[j\,y(n+1) + j\,y(n-1) = x(n+1) - x(n-1) \right] = \tan\omega$$

SOLUTION

Take the DYFT for the lift side

$$\left[j\,y(n+1) + j\,y(n-1) = x(n+1) - x(n-1) \right]$$

$$\left[j\,y(n+1) + j\,y(n-1) = x(n+1) - x(n-1) \right]$$

$$\left[je^{j\omega} + je^{-j\omega} \right] Y\left(e^{j\omega}\right) = \left[e^{j\omega} - e^{-j\omega} \right] X\left(e^{j\omega}\right)$$

$$\frac{Y\left(e^{j\omega}\right)}{X\left(e^{j\omega}\right)} = \frac{[e^{j\omega} + e^{-j\omega}]}{[je^{j\omega} + je^{-j\omega}]} = \frac{\dfrac{[e^{j\omega} - e^{-j\omega}]}{2j}}{\dfrac{[je^{j\omega} + je^{-j\omega}]}{2}}$$

$$= \frac{\sin\omega}{\cos\omega} = \tan\omega$$

6.9   FREQUENCY RESPONSE IN MATLAB

To find the frequency response of the given LTI system given in difference equation form or transfer function form and to plot the same follow the below.

```
clc;
clear all;
close all;
num=input('enter the numerator vector: ');
den=input('enter the denominator vector: ');
N= input(' enter the number of frequency points: ');
w=0:pi/N:pi;
H=freqz(num,den,w);
figure;
subplot(2,2,1);
plot(w/pi,real(H));
xlabel('\omega/\pi');
ylabel('Amplitude');
title('real part');
subplot(2,2,3);
plot(w/pi,imag(H));
xlabel('\omega/\pi');
ylabel('Amplitude');
title('imag part');
subplot(2,2,2);
plot(w/pi,abs(H));
xlabel('\omega/\pi');
ylabel('Magnitude');
title('Magnitude spectrum');
subplot(2,2,4);
plot(w/pi,angle(H));
```

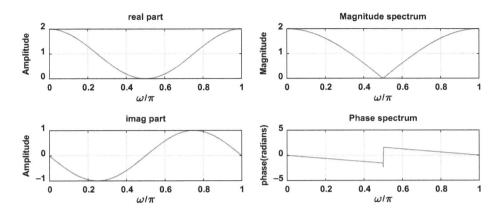

FIGURE 6.6   Frequency response using the MATLAB program.

```
xlabel('\omega/\pi');
ylabel('phase(radians)');
title('Phase spectrum');
```

enter

enter the numerator vector: [1 1 1 1]
enter the denominator vector: [1 1]
enter the number of frequency points: 512

Output

Figure 6.6 shows the frequency response from the output of a MATLAB program.

Example 6.19

Given the *B* and *A* coefficients of a filter as below:

$B = [-0.09355\ -0.01558\ 0.1\ -0.01558\ -0.09355]$
$A = [1]$

Write down a MATLAB program to compute the magnitude frequency response and phase response of this filter. If the sampling rate is 8,000 Hz, then write down MATLAB code to plot the magnitude in dB and phase response in degrees as a function of frequency in Hz.

SOLUTION

The magnitude frequency response and the phase response are computed by the MATLAB program shown below along with the code for the desired plots.

```
[response, w] = freqz([0.1 0.2 0.3 0.2 0.1],[1], 512);
magnitude = abs(response);
magnitude_db = 20*log10(magnitude);
phase_degree = 180*unwrap(angle(response))/pi;
figure,
subplot(2,1,1);
plot(w, magnitude_db);
grid;
xlabel('Frequency (rad)');
ylabel('Magnitude response(dB)');
```

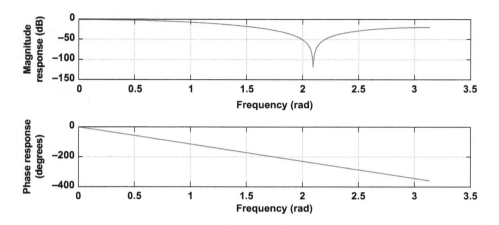

FIGURE 6.7   Frequency response of Example 6.19.

```
subplot(2,1,2);
plot(w, phase_degree);
grid;
xlabel('Frequency (rad)');
ylabel('Phase response(degrees)');
```

Figure 6.7 Shows the frequency response of Example 6.19.

Example 6.20

Given the following digital system with a sampling rate of 8,000 Hz,

$$y(n) = 0.5x(n) + 0.5x(n-1)$$

determine the frequency response of the system.

SOLUTION

Taking the z-transform of both sides of the difference equation, we get

$$Y(z) = 0.5\,X(z) + 0.5\,z^{-1}\,X(z)$$

$$\Rightarrow Y(z) = (0.5 + 0.5\,z^{-1})X(z)$$

Therefore, the transfer function of the system is given by

$$H(z) = \frac{Y(z)}{X(z)} = (0.5 + 0.5\,z^{-1})$$

To find out the frequency response of the system, we replace z with $e^{j\Omega}$, which leads to

$$H\!\left(e^{j\Omega}\right) = \left(0.5 + 0.5\,e^{-j\Omega}\right)$$

The transfer function can be written as

$$H\!\left(e^{j\Omega}\right) = 0.5 + 0.5\cos(\Omega) - j\,0.5\sin(\Omega)$$

Therefore, the magnitude frequency response and phase response are given by

$$\left|H\left(e^{j\Omega}\right)\right| = \left|0.5 + 0.5\cos(\Omega) - j\,0.5\sin(\Omega)\right|$$

$$= \sqrt{\left(0.5 + 0.5\cos(\Omega)\right)^2 + \left(-0.5\sin(\Omega)\right)^2}$$

and

$$\angle H\left(e^{j\Omega}\right) = \tan\left(\frac{-0.5\sin(\Omega)}{0.5 + 0.5\cos(\Omega)}\right)$$

Which is an example of a low-pass filter.

Example 6.21

Given the following digital system with a sampling rate of 8,000 Hz,

$$y(n) = x(n) - 0.5\,y(n-1)$$

determine the frequency response of the system.

SOLUTION

Taking the z-transform of both sides of the difference equation, we get

$$Y(z) = X(z) - 0.5\,z^{-1}Y(z)$$

$$\Rightarrow Y(z) + 0.5\,z^{-1}Y(z) = X(z)$$

$$\Rightarrow \left(1 + 0.5\,z^{-1}\right)Y(z) = X(z)$$

Therefore, the transfer function of the system is given by

$$H(z) = \frac{Y(z)}{X(z)} = \frac{1}{\left(1 + 0.5\,z^{-1}\right)}$$

To find out the frequency response of the system, we replace z with $e^{j\Omega}$, leads to

$$H\left(e^{j\Omega}\right) = \frac{1}{\left(1 + 0.5\,e^{-j\Omega}\right)}$$

The transfer function is written as

$$H\left(e^{j\Omega}\right) = \frac{1}{\left(1 + 0.5\cos(\Omega) - j\,0.5\sin(\Omega)\right)}$$

Therefore, the magnitude frequency response and phase response are given by

$$\left|H\left(e^{j\Omega}\right)\right| = \left|\frac{1}{\left(1 + 0.5\cos(\Omega) - j\,0.5\sin(\Omega)\right)}\right|$$

$$= \frac{1}{\sqrt{\left(1 + 0.5\cos(\Omega)\right)^2 + \left(-0.5\sin(\Omega)\right)^2}}$$

and

$$\angle H\left(e^{j\Omega}\right) = -\tan\left(\frac{-0.5\sin(\Omega)}{1+0.5\cos(\Omega)}\right)$$

This is an example of a high-pass filter.

PROBLEMS

6.1 *Find* the frequency response for the LSI system with unit sample response

$$h(n) = a^n u(n-1)$$

where α is a real number with $|a| < 1$.

6.2 *Find* the frequency response for the LSI system with unit sample response

$$h(n) = a^n \delta(n-1)$$

where α is a real number with $|a| < 1$.

6.3 Find the DTFT of the sequence

$$x(n) = a^n u(n-1) \quad |a| > 1$$

6.4 Find the DTFT of the sequence

$$x(n) = 2^n u(n-1) \quad |a| > 1$$

6.5 Find the inverse DTFT of:

$$X\left(e^{j\omega}\right) = 2\pi\delta(\omega-4) + 2\pi\delta(\omega+4)$$

6.6 Find the frequency response of the linear shift-invariant system characterized by the second-order linear constant-coefficient difference equation

$$y(n) = 1.3433\, y(n-1) - 0.9025\, y(n-2) + x(n) - 1.4142\, x(n-1) + x(n-2)$$

6.7 If the unit sample response of an LSI system is:

$$h(n) = a^n u(n-1)$$

find the response of the system to the input $x(n) = \beta^n u(n)$ where $|a| < 1, |\beta| < 1$ and $a \neq \beta$.

6.8 Solve for $y(n)$ assuming zero initial conditions,

$$y(n) - 0.5\, y(n-1) = x(n) + x(n-1) - x(n-2) \quad \text{for } x(n) = \delta(n).$$

We begin by taking the DTFT of each term in the difference equation:

6.9 Solve for $y(n)$ assuming zero initial conditions,

$$y(n) = x(n) + 0.5x(n-1) - 0.75x(n-2) \quad \text{for } x(n) = \delta(n).$$

We begin by taking the DTFT of each term in the difference equation:

6.10 Find the frequency response for the following in an LSI system
   a) $h(n) = \delta(n) + 0.5\,\delta(n-2) + \delta(n-3)$

   b) $h(n) = \left(\dfrac{1}{5}\right)^{n-3} u(n-3)$

6.11 Find the frequency response for the following in an LSI system

a) $h(n) = \delta(n) + 0.2\,\delta(n-1) + 0.25\,\delta(n-2)$

b) $h(n) = \left(\dfrac{1}{8}\right)^{n+3} u(n+3)$

6.12 Find the DTFT of the following sequence

a) $x_1(n) = (0.75)^n\, u(n-2)$

b) $x_2(n) = a^{n-1} \sin(2n) u(n-1)$

6.13 Find the DTFT of the following sequence

$$y(n) = \begin{cases} (0.5)^{n-1} & \text{for } n \ge 1 \text{ and even only} \\ 0 & \text{otherwise} \end{cases}$$

6.14 Find the DTFT of the following sequence (Figure 6.8).

a) Find the transfer function $H\left(e^{j\omega}\right)$

b) Find the frequency response if

$$h_1(n) = \delta(n+1) + \delta(n-1) + \delta(n-2) = h_3(n)$$

$$h_2(n) = (0.75)^{n+1}\, u(n) = h_4(n)$$

6.15 Find the DTFT of the following sequence (Figure 6.9).

a) Find the transfer function $H\left(e^{j\omega}\right)$

b) Find the frequency response if

$$h_1(n) = \delta(n+1) + 2\,\delta(n) + 3\,\delta(n-2) = h_3(n)$$

$$h_2(n) = (0.9)^{n-1}\, u(n) = h_4(n)$$

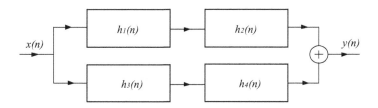

FIGURE 6.8 System of Problem 6.14.

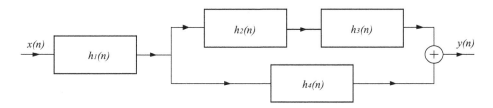

FIGURE 6.9 System of Problem 6.15.

6.16 A LSI system is described by a difference equation

$$y(n) = 0.1\,y(n-2) + \beta x(n)$$

a) Find the gain value α so that the magnitude of the transfer function is equal at dc condition.
b) Find the half power point.

6.17 Find the inverse DTFT

$$X\left(e^{j\omega}\right) = \begin{cases} 5e^{j2\omega} & \dfrac{\pi}{3} < |\omega| < \dfrac{\pi}{3} \\ 0 & \text{otherwise} \end{cases}$$

6.18 Write a program to determine the magnitude and phase responses of the system described by the transfer function:

$$H(S) = \frac{S^2 + \omega_o^2}{S^2 + 2\dfrac{\omega_o}{Q}S + \omega_o^2}$$

Then, run the program and display the results.

6.19 Let $h(n)$ be the unit sample response of a LSI system. Write a program to determine the frequency response when:

$$h(n) = \left(\frac{1}{3}\right)^{n+2} u(n-2)$$

Then, use this program to display the amplitude and phase responses.

6.20 A first-order recursive digital filter is specified by the relation:

$$y(n) = a\,y(n-1) + x(n), \quad \text{for } n \geq 0.$$

where y(n) and x(n) are the output and the input sequences, respectively.
Write a program to determine and display the magnitude and phase responses for this filter.

7 Discrete Fourier Transform (DFT)

Discrete Fourier Transform is one of the great mathematical solutions in DSP to convert the signals from discrete-time to discrete frequency, this chapter will present Method of Decimation-in-Frequency, and Method of Decimation-in-Time for the sequences. The Properties of Discrete Fourier Transform will discuss. Also Discrete Fourier Transform of a sequence in MATLAB and its application in Linear convolution and Generation of Inverse Discrete Fourier Transform (IDFT).

7.1 METHOD OF DECIMATION-IN-FREQUENCY

The DFT of a given sequence $x(n)$, where $n = 0, 1, \ldots, N-1$, can be determined by the following relationship,

$$X(k) = \sum_{n=0}^{N-1} x(n) W_N^{kn} \quad \text{for } k = 0, 1, \ldots, N-1 \tag{7.1}$$

The twiddle factor W_N is given by

$$W_N = e^{-j2\pi/N} = \cos\left(\frac{2\pi}{N}\right) + j\sin\left(\frac{2\pi}{N}\right) \tag{7.2}$$

In this case, N must have the power of two, i.e., it can take up values

$$N = 2, 4, 8, 16, \ldots$$

Equation (7.1) is expanded as

$$X(k) = x(0) + x(1)W_N^k + \cdots + x(N-1)W_N^{k(N-1)} \tag{7.3}$$

The right-hand side of Equation (7.3) can be further split as follows

$$X(k) = x(0) + x(1)W_N^k + \cdots + x\left(\frac{N}{2} - 1\right) W_N^{k\left(\frac{N}{2}-1\right)}$$

$$+ x\left(\frac{N}{2}\right) W_N^{k\left(\frac{N}{2}\right)} + x\left(\frac{N}{2} + 1\right) W_N^{k\left(\frac{N}{2}+1\right)} + \cdots + x(N-1)W_N^{k(N-1)} \tag{7.4}$$

Or written as

$$X(k) = \sum_{n=0}^{(N/2)-1} x(n) W_N^{kn} + \sum_{n=N/2}^{N-1} x(n) W_N^{kn} \tag{7.5}$$

Consider the second summation on the right-hand side of Equation (7.5), i.e.,

$$\sum_{n=N/2}^{N-1} x(n) W_N^{kn}$$

Let us substitute

$$m = n - \frac{N}{2} \Rightarrow \text{at } n = \frac{N}{2}, \ m = 0 \quad \text{and} \quad \text{at } n = N-1, \ m = \frac{N}{2} - 1$$

Therefore,

$$\sum_{n=N/2}^{N-1} x(n)W_N^{kn} = \sum_{m=0}^{(N/2)-1} x\left(m + \frac{N}{2}\right)W_N^{k\left(m + \frac{N}{2}\right)} = W_N^{k\left(\frac{N}{2}\right)} \sum_{m=0}^{(N/2)-1} x\left(m + \frac{N}{2}\right)W_N^{km}$$

As

$$W_N^{N/2} = e^{-j\frac{2\pi}{N}\cdot\frac{N}{2}} = e^{-j\pi} = \cos(\pi) + j\sin(\pi) = -1$$

Therefore,

$$W_N^{k\left(\frac{N}{2}\right)} = \left(W_N^{N/2}\right)^k = (-1)^k$$

Thus,

$$\sum_{n=N/2}^{N-1} x(n)W_N^{kn} = (-1)^k \sum_{m=0}^{(N/2)-1} x\left(m + \frac{N}{2}\right)W_N^{km}$$

As m is a dummy variable (used for summation only), we can use n in its place, i.e.,

$$\sum_{n=N/2}^{N-1} x(n)W_N^{kn} = (-1)^k \sum_{n=0}^{(N/2)-1} x\left(n + \frac{N}{2}\right)W_N^{kn}$$

Thus, the DCT coefficient $X(k)$ (from Equation (7.5)) can be written as

$$X(k) = \sum_{n=0}^{(N/2)-1} x(n)W_N^{kn} + (-1)^k \sum_{n=0}^{(N/2)-1} x\left(n + \frac{N}{2}\right)W_N^{kn} \tag{7.6}$$

For $k = 2m$ as an even number, from Equation (7.6), we can write

$$X(2m) = \sum_{n=0}^{\left(\frac{N}{2}\right)-1} x(n)W_N^{2mn} + (-1)^{2m} \sum_{n=0}^{\left(\frac{N}{2}\right)-1} x\left(n + \frac{N}{2}\right)W_N^{2mn}$$

$$= \sum_{n=0}^{\left(\frac{N}{2}\right)-1} x(n)W_N^{2mn} + \sum_{n=0}^{\left(\frac{N}{2}\right)-1} x\left(n + \frac{N}{2}\right)W_N^{2mn} \tag{7.7}$$

$$= \sum_{n=0}^{\left(\frac{N}{2}\right)-1} \left(x(n) + x\left(n + \frac{N}{2}\right)\right)W_N^{2mn} = \sum_{n=0}^{\left(\frac{N}{2}\right)-1} a(n)W_N^{2mn}$$

And for $k = 2m + 1$ as an odd number, from Equation (7.6), we can write

$$X(2m+1) = \sum_{n=0}^{(N/2)-1} x(n)W_N^{(2m+1)n} + (-1)^{(2m+1)} \sum_{n=0}^{(N/2)-1} x\left(n+\frac{N}{2}\right)W_N^{(2m+1)n}$$

$$= \sum_{n=0}^{(N/2)-1} x(n)W_N^{(2m+1)n} - \sum_{n=0}^{(N/2)-1} x\left(n+\frac{N}{2}\right)W_N^{(2m+1)n}$$

$$= \sum_{n=0}^{(N/2)-1} \left(x(n) - x\left(n+\frac{N}{2}\right)\right)W_N^n W_N^{2mn} = \sum_{n=0}^{(N/2)-1} b(n)W_N^n W_N^{2mn}$$

Using the fact that

$$W_N^2 = e^{-j\frac{2\pi}{N}\cdot 2} = e^{-j\frac{2\pi}{(N/2)}} = W_{N/2}$$

We can write

$$X(2m) = \sum_{n=0}^{(N/2)-1} a(n)W_{N/2}^{mn} = \text{DFT}\left\{a(n) \text{ with } \frac{N}{2} \text{ points}\right\} \tag{7.9}$$

$$X(2m+1) = \sum_{n=0}^{(N/2)-1} b(n)W_N^n W_{N/2}^{mn} = \text{DFT}\left\{b(n)W_N^n \text{ with } \frac{N}{2} \text{ points}\right\} \tag{7.10}$$

where $a(n)$ and $b(n)$ are given by

$$a(n) = x(n) + x\left(n+\frac{N}{2}\right), \quad \text{for } n = 0,1,2,\ldots,\frac{N}{2}-1 \tag{7.11}$$

$$b(n) = x(n) - x\left(n+\frac{N}{2}\right), \quad \text{for } n = 0,1,2,\ldots,\frac{N}{2}-1 \tag{7.12}$$

or collectively

$$\text{DFT}\left\{x(n) \text{ with } N \text{ points}\right\} = \begin{cases} \text{DFT}\left\{a(n) \text{ with } \dfrac{N}{2} \text{ points}\right\} \\ \text{DFT}\left\{b(n)W_N^n \text{ with } \dfrac{N}{2} \text{ points}\right\} \end{cases} \tag{7.13}$$

It could be noted that the number of complex multiplications is drastically reduced (especially for large N) when using FFT instead of DFT. For a sequence with N-points,

$$\text{Number of complex multiplications in DFT} = N^2 \tag{7.14}$$

$$\text{Number of complex multiplications in DFT} = \frac{N}{2}\log_2(N) \tag{7.15}$$

A similar approach has been developed for finding out the inverse DFT using the decimation-in-frequency method of FFT. For the inverse DFT, we have

$$x(n) = \frac{1}{N}\sum_{n=0}^{N-1} X(k)W_N^{-kn} \quad \text{For } n = 0,1,\ldots,N-1 \tag{7.16}$$

This equation is written as,

$$x(n) = \frac{1}{N}\sum_{n=0}^{N-1} X(k)\tilde{W}_N^{kn} \quad \text{For } n = 0,1,\ldots,N-1 \tag{7.17}$$

Whereas,

$$\tilde{W}_N = W_N^{-1} = e^{j\frac{2\pi}{N}} \tag{7.18}$$

Therefore, in this case

$$iDFT\{x(n) \text{ with } N \text{ points}\} = \begin{cases} iDFT\left\{a(n) \text{ with } \frac{N}{2} \text{ points}\right\} \\ iDFT\left\{b(n)\tilde{W}_N^n \text{ with } \frac{N}{2} \text{ points}\right\} \end{cases} \tag{7.19}$$

where $a(n)$ and $b(n)$ are given by

$$a(n) = x(n) + x\left(n + \frac{N}{2}\right), \quad \text{for } n = 0,1,2,\ldots,\frac{N}{2} - 1 \tag{7.20}$$

$$b(n) = x(n) - x\left(n + \frac{N}{2}\right), \quad \text{for } n = 0,1,2,\ldots,\frac{N}{2} - 1 \tag{7.21}$$

7.2   METHOD OF DECIMATION-IN-TIME

In this method, we split the input sequence into the even indexed and odd indexed sequences, as below

$$X(k) = \sum_{n=0}^{N-1} x(n)W_N^{kn} = \sum_{m=0}^{(N/2)-1} x(2m)W_N^{2km} + \sum_{m=0}^{(N/2)-1} x(2m+1)W_N^k \, W_N^{2km} \quad \text{for } k = 0,1,\ldots,N-1$$

whereas,

$$W_N = e^{-j2\pi/N} = \cos\left(\frac{2\pi}{N}\right) + j\sin\left(\frac{2\pi}{N}\right) \tag{7.22}$$

and

$$N = 2,4,8,16,\ldots$$

Using the fact that

$$W_N^2 = e^{-j\frac{2\pi}{N}.2} = e^{-j\frac{2\pi}{(N/2)}} = W_{N/2} \tag{7.23}$$

we get

$$X(k) = \sum_{m=0}^{(N/2)-1} x(2m)W_{N/2}^{km} + W_N^k \sum_{m=0}^{(N/2)-1} x(2m+1)W_{N/2}^{km} \tag{7.24}$$

We define new functions

$$G(k) = \sum_{m=0}^{(N/2)-1} x(2m)W_{N/2}^{km} = \mathrm{DFT}\left\{x(2m) \text{ with } (N/2) \text{ points}\right\} \qquad (7.25)$$

$$H(k) = \sum_{m=0}^{(N/2)-1} x(2m+1)W_{N/2}^{km} = \mathrm{DFT}\left\{x(2m+1) \text{ with } (N/2) \text{ points}\right\} \qquad (7.26)$$

We also note that

$$G(k) = G\left(k + \frac{N}{2}\right), \quad \text{for } k = 0,1,\ldots,\frac{N}{2}-1 \qquad (7.27)$$

$$H(k) = H\left(k + \frac{N}{2}\right), \quad \text{for } k = 0,1,\ldots,\frac{N}{2}-1 \qquad (7.28)$$

So we get

$$X(k) = G(k) + W_N^k H(k), \quad \text{for } k = 0,1,\ldots,\frac{N}{2}-1 \qquad (7.29)$$

Also, keep in view

$$W_N^{\left(\frac{N}{2}+k\right)} = -W_N^k \qquad (7.30)$$

For the second half, we can write

$$X\left(k + \frac{N}{2}\right) = G(k) - W_N^k H(k), \quad \text{for } k = 0,1,\ldots,\frac{N}{2}-1 \qquad (7.31)$$

Let $x(n)$ be a finite duration sequence. Then the N-point DFT of the sequence $x(n)$ is expressed as

$$X[k] = \sum_{k=0}^{N-1} x(n)e^{-\frac{jk2\pi n}{N}} \qquad \mathrm{DFT} \qquad (7.32)$$

where

$$k = 0,1,\ldots,N-1$$

and the corresponding Inverse Discrete Fourier Transform (IDFT) is expressed as

$$x[n] = \frac{1}{N}\sum_{k=0}^{N-1} X(k)e^{\frac{jk2\pi n}{N}} \qquad \mathrm{IDFT} \qquad (7.33)$$

Where

$$n = 0,1,\ldots,N-1$$

Now, let us define the *twiddle factor* as

$$W_n = e^{-\frac{jk2\pi}{N}} \qquad (7.34)$$

Hence, DFT and IDFT equations are written as

$$X(k) = \sum_{n=0}^{N-1} x(n) \, W_N^{kn} \quad k = 0, 1, \ldots, N-1 \tag{7.35}$$

and

$$x(n) = \frac{1}{N} \sum_{k=0}^{N-1} X(k) \, W_N^{-kn} \quad n = 0, 1, \ldots, N-1 \tag{7.36}$$

Example 7.1

Find the DFT of the sequence $x(n) = [1,1,0,0]$ and the IDFT of $Y(k) = [1,0,1,0]$.

SOLUTION

We know that DFT expressed as

$$X[k] = \sum_{k=0}^{N-1} x(n) e^{-\frac{jk2\pi n}{N}}$$

Here $N = 4$, therefore, we have

$$X(k) = \sum_{n=0}^{3} x(n) \, e^{-j\pi nk/2}$$

So when

$$k = 0 \Rightarrow X(0) = \sum_{n=0}^{3} x(n) e^{0}$$

$$= x(0) + x(1) + x(2) + x(3)$$

$$= 1 + 1 + 0 + 0 = 2$$

when

$$k = 1 \Rightarrow X(1) = \sum_{n=0}^{3} x(n) e^{-j\pi n/2}$$

$$= x(0) + x(1)e^{-j\pi/2} + x(2)e^{-j\pi} + x(3)e^{-j3\pi/2}$$

$$= 1 + (1)e^{-j\pi/2} + (0)e^{-j\pi} + (0)e^{-j3\pi/2} = 1 - j$$

when

$$k = 2 \Rightarrow X(2) = \sum_{n=0}^{3} x(n) e^{-j\pi n}$$

$$= x(0) + x(1)e^{-j\pi} + x(2)e^{-j2\pi} + x(3)e^{-j3\pi}$$

$$= 1 + (1)e^{-j\pi} + (0)e^{-j2\pi} + (0)e^{-j3\pi} = 0$$

when

$$k = 3 \Rightarrow X(3) = \sum_{n=0}^{3} x(n) \, e^{-j3\pi n/2}$$

$$= x(0) + x(1)e^{-j3\pi/2} + x(2)e^{-j3\pi} + x(3)e^{-j9\pi/2}$$

$$= 1 + (1)e^{-j3\pi/2} + (0)e^{-j3\pi} + (0)e^{-j9\pi/2} = 1 + j$$

So, $X(k) = [2, 1-j, 0, 1+j]$

Also, we know that the IDFT is expressed as

$$y(n) = \frac{1}{N} \sum_{k=0}^{N-1} Y(k) \, e^{j2\pi nk/N}$$

Here $N = 4$, therefore, we have

$$y(n) = \frac{1}{4} \sum_{m=0}^{3} Y(k) \, e^{j\pi nk/2}$$

So when

$$n = 0 \Rightarrow y(0) = \frac{1}{4} \sum_{k=0}^{3} Y(k) \, e^{0}$$

$$= \frac{1}{4}[Y(0) + Y(1) + Y(2) + Y(3)]$$

$$= \frac{1}{4}[1 + 0 + 1 + 0] = \frac{1}{2}$$

when

$$n = 1 \Rightarrow y(1) = \frac{1}{4} \sum_{k=0}^{3} Y(k) \, e^{j\pi k/2}$$

$$= \frac{1}{4}\left[Y(0) + Y(1)e^{j\pi/2} + Y(2)e^{j\pi} + Y(3)e^{j3\pi/2}\right]$$

$$= \frac{1}{4}\left[1 + (0)e^{j\pi/2} + (1)e^{j\pi} + (0)e^{j3\pi/2}\right] = 0$$

when

$$n = 2 \Rightarrow y(2) = \frac{1}{4} \sum_{k=0}^{3} Y(k) \, e^{j\pi k}$$

$$= \frac{1}{4}\left[Y(0) + Y(1)e^{j\pi} + Y(2)e^{j2\pi} + Y(3)e^{j3\pi}\right]$$

$$= \frac{1}{4}\left[1 + (0)e^{j\pi} + (1)e^{j2\pi} + (0)e^{j3\pi}\right] = \frac{1}{2}$$

when

$$n = 3 \Rightarrow y(3) = \frac{1}{4} \sum_{k=0}^{3} Y(k) \, e^{j3\pi k/2}$$

$$= \frac{1}{4}\left[Y(0) + Y(1)e^{j3\pi/2} + Y(2)e^{j3\pi} + Y(3)e^{j9\pi/2}\right]$$

$$= \frac{1}{4}\left[1 + (0)e^{j3\pi/2} + (1)e^{j3\pi} + (0)e^{j9\pi/2}\right] = 0$$

So, $y(n) = \left[\dfrac{1}{2}, 0, \dfrac{1}{2}, 0\right]$

Example 7.2

Given a sequence $x(n)$ for $0 \leq n \leq 3$, where

$$x(0) = 1, x(1) = 2, x(2) = 3, x(3) = 4$$

   a. Evaluate the DFT, i.e., $X(k)$.
   b. Evaluate and sketch the resulting two-sided amplitude spectrum, i.e., A_k.
   c. Evaluate and draw the resulting one-sided amplitude spectrum, i.e., $\bar{A}_k$.

SOLUTION

(a) According to the definition of DFT

$$X(k) = \sum_{n=0}^{N-1} x(n) W_N^{kn}, \quad \text{for } k = 0, 1, 2, \ldots, N-1$$

where

$$W_N = e^{-j2\pi/N}$$

In this case, $N = 4$, therefore the DFT can be written as

$$X(k) = x(0) + x(1)W_4^k + x(2)W_4^{2k} + x(3)W_4^{3k}$$

Therefore,

$$X(0) = x(0) + x(1) + x(2) + x(3) = 1 + 2 + 3 + 4 = 10$$

$$X(1) = x(0) + x(1)W_4^1 + x(2)W_4^2 + x(3)W_4^3$$

$$= x(0) + x(1)e^{-j2\pi/4} + x(2)e^{-j4\pi/4} + x(3)e^{-j6\pi/4}$$

$$= 1 + (2)(-j) + 3(-1) + 4(j) = -2 + j2$$

$$X(2) = x(0) + x(1)W_4^2 + x(2)W_4^4 + x(3)W_4^6$$

$$= x(0) + x(1)e^{-j4\pi/4} + x(2)e^{-j8\pi/4} + x(3)e^{-j12\pi/4}$$

$$= 1 + (2)(-1) + 3(+1) + 4(-1) = -2$$

$$X(3) = x(0) + x(1)W_4^3 + x(2)W_4^6 + x(3)W_4^9$$

$$= x(0) + x(1)e^{-j6\pi/4} + x(2)e^{-j12\pi/4} + x(3)e^{-j18\pi/4}$$

$$= 1 + (2)(j) + 3(-1) + 4(-j) = -2 - j2$$

(b) Two-sided amplitude spectrum
   The two-sided amplitude spectrum is given by

$$A_k = \frac{1}{N}|X(k)| = \frac{1}{N}\sqrt{\left(\text{Real}[X(k)]\right)^2 + \left(\text{Imag}[X(k)]\right)^2}$$

Therefore (Figure 7.1),

$$A_0 = \frac{1}{4}\sqrt{(10)^2 + (0)^2} = \frac{10}{4} = 2.5$$

$$A_1 = \frac{1}{4}\sqrt{(-2)^2 + (2)^2} = \frac{\sqrt{8}}{4} = 0.7071$$

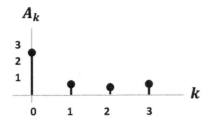

FIGURE 7.1 Amplitude spectrum of Example 7.2 part b.

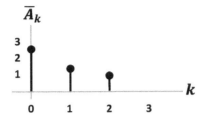

FIGURE 7.2 Amplitude spectrum of Example 7.2 part c.

$$A_2 = \frac{1}{4}\sqrt{(-2)^2 + (0)^2} = \frac{2}{4} = 0.5$$

$$A_3 = \frac{1}{4}\sqrt{(-2)^2 + (-2)^2} = \frac{\sqrt{8}}{4} = 0.7071$$

(c) One-sided amplitude spectrum
The one-sided amplitude spectrum is given by

$$\bar{A}_k = \begin{cases} \dfrac{1}{N}|X(k)| & k = 0 \\ \dfrac{2}{N}|X(k)| & k = 1, 2, \ldots, N/2 \end{cases}$$

Now, $\dfrac{N}{2} = \dfrac{4}{2} = 2$ Therefore (Figure 7.2),

$$\bar{A}_0 = 2.5$$

$$\bar{A}_1 = 2 \times 0.7071 = 1.4141$$

$$\bar{A}_2 = 2 \times 0.5 = 1.0$$

It is observed that since the sequence $x(n)$ did not have any symmetry with the origin, the DFT, $X(k)$ contained complex values even though $x(n)$ is real. In the case of $Y(k)$, having even symmetry, for the origin, both $Y(k)$ and $y(n)$ are real.

7.3 PROPERTIES OF DISCRETE FOURIER TRANSFORM

1) *Periodicity*
This property states that if a discrete-time signal is periodic, then its DFT will also be periodic. Also, if a signal or sequence repeats its waveform after several N samples, then it is called a periodic signal or sequence and N is called the period of the signal. Mathematically,

if $X(k)$ is an N-point DFT of $x(n)$, then we have

$$x(n+N) = x(n) \text{ for all values of } n$$
$$X(k+N) = X(k) \text{ for all values of } k \tag{7.37}$$

2) *Linearity*

The linearity property states that if $X_1(k)$ and $X_2(k)$ are the N-points DFT of $x_1(n)$ and $x_2(n)$ respectively, a and b are arbitrary constants either real or complex-valued, then we have

$$ax_1(n) + bx_2(n) \Leftrightarrow aX_1(k) + bX_2(k) \tag{7.38}$$

3) *Convolution*

If $x_1(n) \; X_1(k)$ and $x_2(n) \; X_2(n)$

Then,

$$\sum_{m=0}^{N-1} x_1(n) \cdot x_2(n) \leftrightarrow X_1(k) \cdot X_2(k) \tag{7.39}$$

Example 7.3

Find the DFT values for the sample data sequence $x(n) = [1, 1, 2, 2, 3, 3]$ and determine the corresponding amplitude and phase spectrum.

SOLUTION

It is known that the N-point DFT of a finite sequence $x(n)$ is defined as

$$X(k) = \sum_{n=0}^{N-1} x(n) e^{-j2\pi nk/N} \quad k = 0, 1, \dots, N-1$$

We have $N = 6$, then $X(k) = \sum_{n=0}^{5} x(n) e^{j\pi nk/3}$

So,

$$X(0) = \sum_{n=0}^{5} x(n) e^{0}$$
$$= x(0) + x(1) + x(2) + x(3) + x(4) + x(5)$$
$$= 1 + 1 + 2 + 2 + 3 + 3 = 12$$

$$X(1) = \sum_{n=0}^{5} x(n) e^{-j\pi n/3}$$

$$= x(0) + x(1)e^{-j\pi/3} + x(2)e^{-j2\pi/3} + x(3)e^{-j\pi} + x(4)e^{-j4\pi/3} + x(5)e^{-j5\pi/3}$$

$$= 1 + (1)e^{-j\pi/3} + (2)e^{-j2\pi/3} + (2)e^{-j\pi} + (3)e^{-j4\pi/3} + (3)e^{-j5\pi/3}$$

$$= 1 + (0.5 - j0.866) + 2(-0.5 - j0.866) + 2(-1) + 3(-0.5 + j0.866) + 3(0.5 + j0.866)$$

$$= -1.5 + j2.598$$

$$X(2) = \sum_{n=0}^{5} x(n) e^{-j2\pi n/3}$$

$$= x(0) + x(1)e^{-j2\pi/3} + x(2)e^{-j4\pi/3} + x(3)e^{-j2\pi} + x(4)e^{-j8\pi/3} + x(5)e^{-j10\pi/3}$$

$$= 1 + (1)e^{-j2\pi/3} + (2)e^{-j4\pi/3} + (2)e^{-j2\pi} + (3)e^{-j8\pi/3} + (3)e^{-j10\pi/3}$$

$$= 1 + (-0.5 - j0.866) + 2(-0.5 + j0.866) + 2(1) + 3(-0.5 - j0.866) + 3(-0.5 + j0.866)$$

$$= -1.5 + j0.866$$

$$X(3) = \sum_{n=0}^{5} x(n) e^{-j2\pi n}$$

$$= x(0) + x(1)e^{-j\pi} + x(2)e^{-j2\pi} + x(3)e^{-j3\pi} + x(4)e^{-j4\pi} + x(5)e^{-j5\pi}$$

$$= 1 + (1)e^{-j\pi} + (2)e^{-j2\pi} + (2)e^{-j3\pi} + (3)e^{-j4\pi} + (3)e^{-j5\pi}$$

$$= 1 - 1 + 2(1) + 2(-1) + 3(1) + 3(-1)$$

$$X(4) = \sum_{n=0}^{5} x(n) e^{-j4\pi n/3}$$

$$= x(0) + x(1)e^{-j4\pi/3} + x(2)e^{-j8\pi/3} + x(3)e^{-j4\pi} + x(4)e^{-j16\pi/3} + x(5)e^{-j20\pi/3}$$

$$= 1 + (1)e^{-j4\pi/3} + (2)e^{-j8\pi/3} + (2)e^{-j4\pi} + (3)e^{-j16\pi/3} + (3)e^{-j20\pi/3}$$

$$= 1 + (-0.5 + j0.866) + 2(-0.5 - j0.866) + 2(1) + 3(-0.5 + j0.866) + 3(-0.5 - j0.866)$$

$$= -1.5 - j0.866$$

$$X(5) = \sum_{n=0}^{5} x(n) e^{-j5\pi n/3}$$

$$= x(0) + x(1)e^{-j5\pi/3} + x(2)e^{-j10\pi/3} + x(3)e^{-j5\pi} + x(4)e^{-j20\pi/3} + x(5)e^{-j25\pi/3}$$

$$= 1 + (1)e^{-j5\pi/3} + (2)e^{-j10\pi/3} + (2)e^{-j5\pi} + (3)e^{-j20\pi/3} + (3)e^{-j25\pi/3}$$

$$= 1 + (0.5 + j0.866) + 2(-0.5 + j0.866) + 2(-1) + 3(-0.5 - j0.866) + 3(0.5 - j0.866)$$

$$= -1.5 - j2.598$$

Hence,

$$X(k) = \begin{bmatrix} 12 & -1.5 + j2.598 & -1.5 + j0.866 & 0 & -1.5 - j0.866 & -1.5 - j2.598 \end{bmatrix}$$

The corresponding amplitude spectrum is found as

$$|X(k)| = [12, 3, 1.732, 0, 1.732, 3]$$

and the corresponding phase spectrum is

$$\angle X(k) = [0°\ 120°\ 150°\ 0°\ -150°\ -120°]$$

Example 7.4

Evaluate the DFT of the sequence $X[n] = [1,0,0,1]$, $n \geq 0$.

SOLUTION

$N = 4 \ (k = 0 \rightarrow 4-1 = 3)$

$$X[k] = \sum_{n=0}^{3} x(n) e^{-\frac{jk2\pi n}{4}}$$

$$X[0] = x[0] + x[1] + x[2] + x[3] = 2$$

$$X[1] = 1 + e^{-\frac{j2\pi}{2}} = 1 + j$$

$$X[2] = 0$$

$$X[1] = 1 - j$$

Example 7.5

Find the circular convolution between $x_1(n) = [1,1,2,2]$ and $x_2(n) = [1,2,3,4]$ using DFT and IDFT.

SOLUTION

For $x_1(n) = [1, 1, 2, 2]$, we have

$$X_1(k) = \sum_{n=0}^{N-1} x_1(n) e^{-j2\pi nk/N}$$

Here, $N = 4$, therefore, we have

$$X_1(k) = \sum_{n=0}^{3} x_1(n) e^{-j\pi nk/2}$$

So, when

$$k = 0 \Rightarrow X_1(0) = \sum_{n=0}^{3} x_1(n) e^{0}$$

$$= x(0) + x(1) + x(2) + x(3)$$

$$= 1 + 1 + 2 + 2 = 6$$

when

$$k = 1 \Rightarrow X_1(1) = \sum_{n=0}^{3} x_1(n) e^{-j\pi n/2}$$

$$= x_1(0) + x_1(1)e^{-j\pi/2} + x_1(2)e^{-j\pi} + x_1(3)e^{-j3\pi/2}$$

$$= 1 + (1)e^{-j\pi/2} + (2)e^{-j\pi} + (2)e^{-j3\pi/2}$$

$$= 1 + (0 - j) + 2(-1 - j0) + 2(0 + j) = -1 + j$$

when

$$k = 2 \Rightarrow X_1(2) = \sum_{n=0}^{3} x_1(n) e^{-j\pi n}$$

$$= x_1(0) + x_1(1)e^{-j\pi} + x_1(2)e^{-j2\pi} + x_1(3)e^{-j3\pi}$$

$$= 1 + (1)e^{-j\pi} + (2)e^{-j2\pi} + (2)e^{-j3\pi}$$

$$= 1 + (-1 - j0) + 2(1 - j0) + 2(-1 - j0) = 0$$

when

$$k = 3 \Rightarrow X_1(3) = \sum_{n=0}^{3} x_1(n)\, e^{-j3\pi n/2}$$

$$= x_1(0) + x_1(1)e^{-j3\pi/2} + x_1(2)e^{-j3\pi} + x_1(3)e^{-j9\pi/2}$$

$$= 1 + (1)e^{-j3\pi/2} + (2)e^{-j3\pi} + (2)e^{-j9\pi/2}$$

$$= 1 + (0+j) + 2(-1-j0) + 2(0-j) = -1-j$$

So, $X_1(k) = [6, -1+j, 0, -1-j]$

For $x_2(n) = [1, 2, 3, 4]$, we have $X_2(k) = \sum_{n=0}^{3} x_2(n)\, e^{-j\pi nk/2}$
So, when

$$k = 0 \Rightarrow X_2(0) = \sum_{n=0}^{3} x_2(n)\, e^{0}$$

$$= x_2(0) + x_2(1) + x_2(2) + x_2(3)$$

$$= 1 + 2 + 3 + 4 = 10$$

when

$$k = 1 \Rightarrow X_2(1) = \sum_{n=0}^{3} x_2(n)\, e^{-j\pi n/2}$$

$$= x_2(0) + x_2(1)e^{-j\pi/2} + x_2(2)e^{-j\pi} + x_2(3)e^{-j3\pi/2}$$

$$= 1 + (2)e^{-j\pi/2} + (3)e^{-j\pi} + (4)e^{-j3\pi/2}$$

$$= 1 + 2(0-j) + 3(-1-j0) + 4(0+j) = -2 + j2$$

when

$$k = 2 \Rightarrow X_2(2) = \sum_{n=0}^{3} x_2(n)\, e^{-j\pi n}$$

$$= x_2(0) + x_2(1)e^{-j\pi} + x_2(2)e^{-j2\pi} + x_2(3)e^{-j3\pi}$$

$$= 1 + (2)e^{-j\pi} + (3)e^{-j2\pi} + (4)e^{-j3\pi}$$

$$= 1 + 2(-1-j0) + 3(1-j0) + 4(-1-j0) = -2$$

when

$$k = 3 \Rightarrow X_2(3) = \sum_{n=0}^{3} x_2(n)\, e^{-j3\pi n/2}$$

$$= x_2(0) + x_2(1)e^{-j3\pi/2} + x_2(2)e^{-j3\pi} + x_2(3)e^{-j9\pi/2}$$

$$= 1 + (2)e^{-j3\pi/2} + (3)e^{-j3\pi} + (4)e^{-j9\pi/2}$$

$$= 1 + 2(0+j) + 3(-1-j0) + 4(0-j) = -2 - j2$$

So,

$$X_2(k) = [10, -2+j2, -2, -2-j2]$$

Then

$$X_3(k) = X_1(k) \cdot X_2(k)$$

$$= [60, (-1+j)(-2+j2), 0, (-1-j)(-2-j2)]$$

$$= [60, -4j, 0, 4j]$$

We know that $x_3(n) = \text{IDFT}\{X_3(k)\}$

$$x_3(n) = \frac{1}{4}\sum_{m=0}^{3} X_3(k)\, e^{j\pi nk/2} \quad n = 0, 1, 2, 3$$

$$= \frac{1}{4}\left[60 - j4e^{jn\pi/2} + j4e^{j3n\pi/2}\right]$$

when

$$n = 0 \Rightarrow x_3(0) = \frac{1}{4}\left[60 - j4 + j4\right] = 15$$

when

$$n = 1 \Rightarrow x_3(1) = \frac{1}{4}\left[60 - j4e^{j\pi/2} + j4e^{j3\pi/2}\right]$$

$$= \frac{1}{4}\left[60 - j4(j) + j4(-j)\right]$$

$$= \frac{1}{4}\left[60 + 4 + 4\right] = 17$$

when

$$n = 2 \Rightarrow x_3(2) = \frac{1}{4}\left[60 - j4e^{j\pi} + j4e^{j3\pi}\right]$$

$$= \frac{1}{4}\left[60 - j4(-1) + j4(-1)\right]$$

$$= \frac{1}{4}\left[60 + j4 - j4\right] = 15$$

when

$$n = 3 \Rightarrow x_3(3) = \frac{1}{4}\left[60 - j4e^{j3\pi/2} + j4e^{j9\pi/2}\right]$$

$$= \frac{1}{4}\left[60 - j4(-j) + j4(j)\right]$$

$$= \frac{1}{4}\left[60 - 4 - 4\right] = 13$$

Therefore, $x_3(n) = [15, 17, 15, 13]$.

7.4   DISCRETE FOURIER TRANSFORM OF A SEQUENCE IN MATLAB

To find the Discrete Fourier Transform of a sequence use the below.

```
clc;
clear all;
close all;
M=input('Enter the sequence :');
```

```
N=length(M);
for k=1:N
y(k)=0;
for i=1:N
y(k)=y(k)+M(i)*exp((-2*pi*j/N)*((i-1)*(k-1)));
end;
end;
k=1:N
disp('The result is:');y
figure(1);
subplot(211);
stem(k,abs(y(k)));
grid;
xlabel('n');
ylabel('Amplitudes');
title('Magnitude response');
subplot(212);
stem(angle(y(k))*180/pi);
grid;
xlabel('n');
ylabel('phase');
title('Phase response');
```

Enter the sequence :[2 −1 −2 4]
$k = 1 \quad 2 \quad 3 \quad 4$

The result is:
$y = [3.0000 + 0.0000i \; 4.0000 + 5.0000i - 3.0000 - 0.0000i]$

Figure 7.3 shows an amplitude and phase spectrum using DFT of a sequence in MATLAB.

7.5 LINEAR CONVOLUTION USING THE DFT

The DFT provides a convenient way to perform convolutions without having to evaluate the convolution sum. Specifically, if $h(n)$ is N_1 points long, and $x(n)$ is N_2 points long, $h(n)$ may be linearly convolved with $x(n)$ as follows:

1. Pad the sequences $h(n)$ and $x(n)$ with zeros so that they are of the length $N \geq N_1 + N_2 - 1$.
2. Find the N-point DFTs of $h(n)$ and $x(n)$.

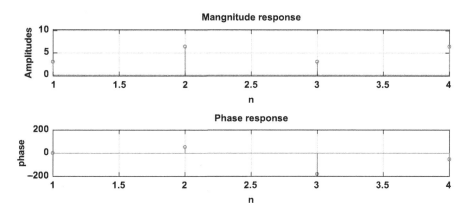

FIGURE 7.3 Amplitude and phase spectrum using DFT of a sequence in MATLAB.

3. Multiply the DFTs to form the product $Y(k) = H(k) X(k)$.
4. Find the inverse DFT of $Y(k)$.

7.6   GENERATION OF INVERSE DISCRETE FOURIER TRANSFORM IN MATLAB

To find the Inverse Discrete Fourier Transform of a sequence use the below.

```
clc;
clear all;
close all;
M=input('Enter the sequence :');
N=length(M);
for n=1:N
y(n)=0;
for k=1:N
y(n)=y(n)+M(k)*exp((2*pi*j*(k-1)*(n-1))/N);
end;
end;
n=1:N
x=1/N*y(n)
figure(1);
stem(n,x);
grid;
xlabel('n');
ylabel('Amplitudes');
title('Mangnitude response');
```

Enter the sequence: [1 −2 −3 4]
$n = 1$ 2   3   4
$x = [0.0000 + 0.0000i\ 1.0000 − 1.5000i\ −1.0000 + 0.0000i\ 1.0000 + 1.5000i]$

Figure 7.4 shows an amplitude spectrum using IDFT of a sequence in MATLAB.

Example 7.6

Evaluate the IDFT of the sequence $X[k] = [2,1 + j,0,1 − j]$

$$x[n] = \frac{1}{N} \sum_{k=0}^{N-1} X(k) e^{\frac{jk2\pi n}{N}}$$

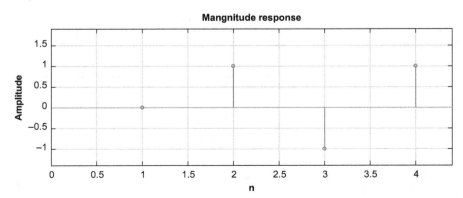

FIGURE 7.4   Amplitude spectrum using IDFT of a sequence in MATLAB.

$$x[n] = \frac{1}{4} \sum_{k=0}^{3} X(k) e^{\frac{jk \cdot 2\pi n}{4}}$$

$$x[0] = \frac{1}{4} \sum_{k=0}^{3} X(k) e^{\frac{jk 2\pi 0}{4}} = 1$$

$$x[1] = \frac{1}{4} \sum_{k=0}^{3} X(k) e^{\frac{jk 2\pi 1}{4}} = 0$$

$$x[2] = \frac{1}{4} \sum_{k=0}^{3} X(k) e^{\frac{jk 2\pi 2}{4}} = 0$$

$$x[3] = \frac{1}{4} \sum_{k=0}^{3} X(k) e^{\frac{jk 2\pi 3}{4}} = 1$$

So x[n]=[1 0 0 1]

Example 7.7

Given a discrete-time sequence $x(n)$ = [1 2 3 4 5], use the MATLAB program to estimate the tone frequency using DFT.

SOLUTION

```
clc;
clear all;
N=5;
x=[1 2 3 4 5];
subplot(2,1,1);
Y=abs(fft(x))
stem([0:N-1],abs(fft(x)));
  xlabel('k');
ylabel('amplitude');
  title('Magnitude Response');
  grid
  subplot(2,1,2);
  stem([0:N-1],angle(fft(x)));
  xlabel('k');
ylabel('amplitude');
  title('Phase Response');
grid
```

Y = [15.0000 4.2533 2.6287 2.6287 4.2533]

Figure 7.5 shows an amplitude and phase spectrum using DFT for Example 7.7.

PROBLEMS

7.1   Find the DFT of the sequence $x(n) = [1,1,1,1]$.
7.2   Find the IDFT of the sequence $x(n) = [1,1,1,1]$.

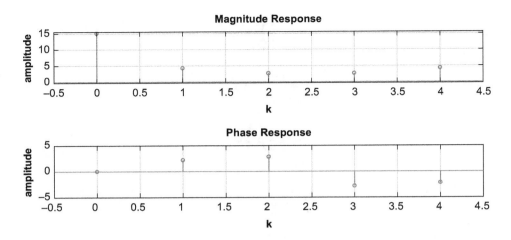

FIGURE 7.5   Amplitude and phase spectrum using DFT of Example 7.7.

7.3   Given a sequence $x(n)$ for $0 \le n \le 3$, where

$$x(0) = 4, \quad x(1) = 3, \quad x(2) = 2, \quad x(3) = 1$$

   a. Evaluate its DFT, i.e., $X(k)$.
   b. Evaluate and sketch the resulting two-sided amplitude spectrum, i.e., A_k.
   c. Evaluate and draw the resulting one-sided amplitude spectrum, i.e., $\bar{A}_k$.

7.4   Given a sequence $x(n)$ for $0 \le n \le 3$, where

$$x(0) = 1, \quad x(1) = -1, \quad x(2) = -1, \quad x(3) = 1$$

   a. Evaluate its DFT, i.e., $X(k)$.
   b. Evaluate and sketch the resulting two-sided amplitude spectrum, i.e., A_k.
   c. Evaluate and draw the resulting one-sided amplitude spectrum, i.e., $\bar{A}_k$.

7.5   Find the DFT values for the sample data sequence $x(n) = [1,-1,2,-2,3,-3]$ and determine the corresponding amplitude and phase spectrum.

7.6   Find the DFT values for the sample data sequence $x(n) = [1,-1,1,-1,1,-1]$ and determine the corresponding amplitude and phase spectrum.

7.7   Evaluate the DFT of the sequence $X[n] = [-1,0,0,1]$, $n \ge 1$.

7.8   Find the circular convolution between $x_1(n) = [1,-1,-2,2]$ and $x_2(n) = [1,2,3,4]$ using DFT and IDFT.

7.9   Find the circular convolution between $x_1(n) = [1,1,-2,-2]$ and $x_2(n) = [1,2,3,4]$ using DFT and IDFT.

7.10   Given a sequence $x(n)$ for $0 \le n \le 3$, where $x(0) = 1$, $x(1) = 2$, $x(2) = 3$, and $x(3) = 4$, calculate DFT $X(k)$

7.11   Given a sequence $x(n)$ for $0 \le n \le 3$, where $x(0) = 1$, $x(1) = 1$, $x(2) = -1$, and $x(3) = 0$, calculate DFT $X(k)$

7.12   Write a program to find the DFT of $h(n)$ and $x(n)$. Then, perform the linear convolution of the two sequences below in the time domain. Compare with the result obtained by using the DFT relations.

$$h(n) = \begin{cases} 0.5 & \text{for } n = 0 \\ 1 & \text{for } n = 1 \\ 0 & \text{otherwise} \end{cases} \quad \text{and} \quad x(n) = \begin{cases} 1 & \text{for } n = 0 \\ 0.5 & \text{for } n = 1 \\ 0 & \text{otherwise} \end{cases}$$

7.13 A signal has the following signal values $x(n) = [1,2,0,3]$, estimate the real and imaginary parts using:

I-   DFT.

II- DFS.

III- DTFT.

Write a MATLAB program to find the DFT and IDFT convolution of $h(n)$ and $x(n)$ by using the matrix method.

$$h(n) = \sin(n * \pi / 4) * \big[u(n) - u(n - 4)\big] * 2^{n-1}$$

$$x(n) = \cos(n * \pi / 4) * \big[u(n - 1) - u(n - 5)\big]$$

7.14 Let $x(n)$ be the sequence:

$$x(n) = 2\delta(n) + \delta(n - 2) + \delta(n - 4)$$

The five-point DFT of $x(n)$ is computed, and the resulting sequence is squared:

$$Y(k) = X_2(k)$$

A five-point inverse DFT is then computed to produce the sequence $y(n)$. Write a program to find the sequence $y(n)$.

7.15 Consider a digital sequence sampled at the rate of 20,000 Hz. If we use the 8,000-point DFT to compute the spectrum, determine:

a. The frequency resolution.

b. The folding frequency in the spectrum.

7.16 Use the DFT to compute the amplitude spectrum of a sampled data sequence with a sampling rate fs = 2,000 Hz. It requires the frequency resolution to be less than 0.5 Hz. Determine the number of data points used by the FFT algorithm and actual frequency resolution in Hz, if the data samples are available for selecting the number of data points.

8 Fast Fourier Transform (FFT)

FFT is a very efficient method for computing the DFT coefficients. It reduces the number of complex multiplications from N^2 in case of DFT to simply $(N/2)\log_2(N)$ In the case of FFT. The only restriction on the algorithm is that the sequence $x(ne)$ should consist of 2^m Samples, where m is a positive integer – in other words, the number N of samples in the sequence should be a power of 2, i.e., $N = 2, 4, 8, 16, \ldots$ etc. If $x(n)$ does not contain 2^m samples, then we append it with zeros until the number of samples in the resulting sequence becomes a power of 2.

There are several ways in which the FFT could). We shall study radix-2 FFT algorithms, namely,

- Decimation-in-Frequency method.
- Decimation-in-Time method.

Other types include radix-4 and split-radix methods.

8.1 FAST FOURIER TRANSFORM DEFINITION

The Fast Fourier Transform (FFT) is defined as an algorithm which efficiently computes the Discrete Fourier Transform (DFT). It is worth mentioning that FFT gives the correct results of DFT values, not an approximation. It is a DFT with a reduced number of necessary arithmetic operations. Table 8.1 shows the principle truth table as a base to build up the FFT algorithm where the binary number represent the input to the FFT and the reversed binary represent the output of the FFT.

8.1.1 Decimation-in-Time FFT

The *shuffling* of the input sequence that takes place is due to the successive decimations of $x(n)$. A complete eight-point decimation-in-time FFT is shown in Figure 8.1. The ordering that the results correspond to are a bit-reversed indexing of the original sequence. In other words, if the index n was written in binary form, the order in which in the input sequence must be accessed is found by reading the binary representation for n in reverse order as illustrated in the Table 8.1 for $N = 8$. Finally, note that the input sequence $x(n)$ is in bit-reversed order, and the frequency samples $X(k)$ are in normal order.

8.1.2 Decimation-in-Frequency FFT

Another class of FFT algorithms is the decimation-in-frequency. A complete eight-point decimation-in-frequency FFT is shown in Figure 8.2. The complexity of the decimation-in-frequency FFT is the same as the decimation-in-time, and the computations performed in place. Finally, note that the input sequence $x(n)$ is in normal order, and the frequency samples $X(k)$ are in bit-reversed order (Figure 8.2).

TABLE 8.1
FFT Algorithm Truth Table

n	Binary	Bit-reversed binary	n'
0	000	000	0
1	001	100	4
2	010	010	2
3	011	110	6
4	100	001	1
5	101	101	5
6	110	011	3
7	111	111	7

Example 8.1

Find the decimation-in-time FFT for the data sequence $x(0) = 0.3535$, $x(1) = 0.3535$, $x(2) = 0.6464$, $x(3) = 1.0607$, $x(4) = 0.3535$, $x(5) = -1.0607$, $x(6) = -1.3535$, $x(7) = -0.3535$.

SOLUTION

By applying the input values, the output of the second stage will be

$$A(0) = 0.707 + W_4^0(-0.707) = 0.707 + (1 + j0)(-0.707) = 0 + j0$$

$$A(1) = 0 + W_4^1(1.999) = 0 + (0 - j1)(1.999) = 0 - j1.999$$

$$A(2) = 0.707 + W_4^2(-0.707) = 0.707 + (-1 + j0)(-0.707) = 1.414 + j0$$

$$A(3) = 0 + W_4^3(1.999) = 0 + (0 + j1)(1.999) = 0 + j1.999$$

$$B(0) = -0.707 + W_4^0(0.707) = -0.707 + (1 + j0)(0.707) = 0 + j0$$

$$B(1) = 1.414 + W_4^1(1.414) = 1.414 + (0 - j1)(1.414) = 1.414 - j1.414$$

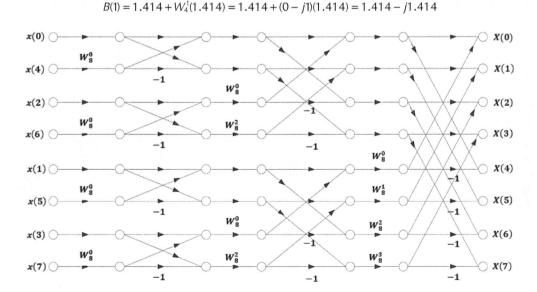

FIGURE 8.1   Eight-point decimation-in-time FFT.

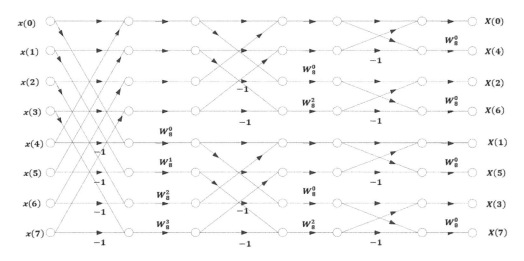

FIGURE 8.2   Eight-point decimation-in-frequency FFT.

$$B(2) = -0.707 + W_4^2(0.707) = -0.707 + (-1 + j0)(0.707) = -1.414 + j0$$

$$B(3) = 1.414 + W_4^3(1.414) = 1.414 + (0 + j1)(1.414) = 1.414 + j1.414$$

Calculate the outputs of the third stage of the FFT to arrive at the final answer

$$X(0) = A(0) + W_8^0 B(0) = 0 + j0 + (1 - j0)(0 + j0) = 0 + j0 + 0 + j0 = 0\angle 0°$$

$$X(1) = A(1) + W_8^1 B(1) = 0 - j1.999 + (0.707 - j0.707)(1.414 - j1.414)$$

$$= 0 - j1.999 + 0 - j1.999 = 0 - j4 = 4\angle -90°$$

$$X(2) = A(2) + W_8^2 B(2) = 1.414 + j0 + (0 - j1)(-1.414 + j0)$$

$$= 1.414 + j0 + 0 + j1.414 = 1.414 + j1.414 = 2\angle 45°$$

$$X(3) = A(3) + W_8^3 B(3) = 0 + j1.999 + (-0.707 - j0.707)(1.414 + j1.414)$$

$$= 0 + j1.999 + 0 - j1.999 = 0\angle 0°$$

$$X(4) = A(0) + W_8^4 B(0) = 0 + j0 + (-1 + j0)(0 + j0) = 0 + j0 + 0 + j0 = 0\angle 0°$$

$$X(5) = A(1) + W_8^5 B(1) = 0 - j1.999 + (-0.707 + j0.707)(1.414 - j1.414)$$

$$= 0 - j1.999 + 0 + j1.999 = 0 - j0 = 0\angle 0°$$

$$X(6) = A(2) + W_8^6 B(2) = 1.414 + j0 + (0 + j1)(-1.414 + j0)$$

$$= 1.414 + j0 + 0 - j1.414 = 1.414 - j1.414 = 2\angle -45°$$

$$X(7) = A(3) + W_8^7 B(3) = 0 + j1.999 + (0.707 + j0.707)(1.414 + j1.414)$$

$$= 0 + j1.999 + 0 + j1.999 = 0 + j4 = 4\angle 90°$$

Figure 8.3 shows the *eight-point decimation-in-time FFT.*

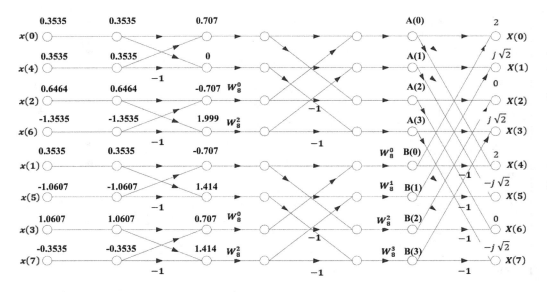

FIGURE 8.3   Eight-point decimation-in-time FFT of Example 8.1.

Example 8.2

Find the decimation-in-time FFT for the data sequence

$$x(n) = \begin{bmatrix} 0.5 & 0 & 0.5 & 0 & 0.5 & 0 & 0.5 & 0 \end{bmatrix}.$$

SOLUTION

Figure 8.4 shows the eight-point decimation-in-time FFT.

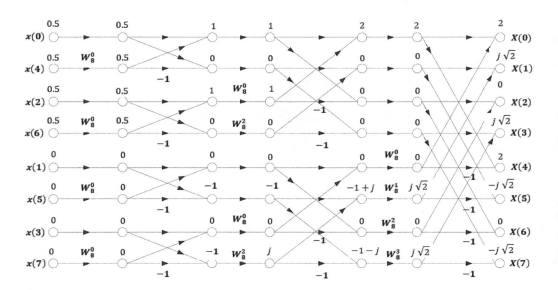

FIGURE 8.4   Eight-point decimation-in-time FFT of Example 8.2.

Example 8.3

Given a sequence $x(n)$ For $0 \leq n \leq 3$, where $x(0) = 1$, $x(1) = 2$, $x(2) = 3$, and $x(3) = 4$:

    a. Determine its DFT $X(k)$ using the decimation-in-frequency FFT method?
    b. Determine the number of complex multiplications by doing part (a).

SOLUTION

(a) According to the FFT (decimation-in-frequency method) for calculating DFT of a given sequence

$$DFT\left\{x(n) \text{ with } N \text{ points}\right\} = \begin{cases} DFT\left\{a(n) \text{ with } \dfrac{N}{2} \text{ points}\right\} \\ DFT\left\{b(n) \, W_N^n \text{ with } \dfrac{N}{2} \text{ points}\right\} \end{cases}$$

where, $W_N^n = e^{-j2\pi n/N} = \cos\left(\dfrac{2\pi n}{N}\right) + j\sin\left(\dfrac{2\pi n}{N}\right)$
and

$$a(n) = x(n) + x\left(n + \frac{N}{2}\right), \quad \text{for} \quad n = 0, 1, 2, \dots, \frac{N}{2} - 1$$

$$b(n) = x(n) - x\left(n + \frac{N}{2}\right), \quad \text{for} \quad n = 0, 1, 2, \dots, \frac{N}{2} - 1$$

The DFT of the given sequence, using FFT (decimation-in-frequency method) is given in the following Figures 8.5 and 8.6, respectively.
Therefore, the DFT coefficients are $X(0) = 10$, $X(1) = -2 + 2j$, $X(2) = -2$, and $X(3) = -2 - 2j$.

(b) Number of complex multiplications in $DFT = \dfrac{N}{2}\log_2(N) = \dfrac{4}{2}\log_2(4) = 4$.

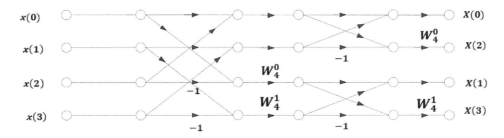

FIGURE 8.5 Four-point decimation-in-time FFT.

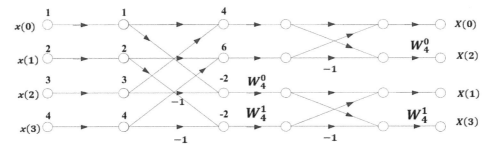

FIGURE 8.6 Four-point decimation-in-time FFT of Example 8.3.

8.2 FINDING THE FFT OF DIFFERENT SIGNALS IN MATLAB

To find the FFT of different signals like an impulse, step, ramp, and exponential. The FFT can generate these by using the command FFT for impulse, step, ramp, and exponential sequences. In the process of finding the FFT, the length of the FFT is taken as *N*. The FFT consists of two parts: magnitude plot and phase plot. The magnitude plot is the absolute value of magnitude versus the samples, and the phase plot is the phase angle versus the samples.

```
clc;
clear all;
close all;
t=-2:1:2;
x=[zeros(1,2) 1 zeros(1,2)];
subplot (3,1,1);
stem(t,x);
grid;
x=[0 0 1 0 0];
disp(x);
title ('Impulse Response');
xlabel ('n');
ylabel ('Amplitude');
yn=x;
N=30;
xk=fft(yn,N);
magxk=abs(xk);
angxk=angle(xk);
k=0:N-1;
subplot(3,1,2);
stem(k,magxk);
grid;
xlabel('k');
ylabel('|y(k)|');
subplot(3,1,3);
stem(k,angxk);
disp(xk);
grid;
xlabel('k');
ylabel('arg(y(k))');
```

Figure 8.7 shows the impulse FFT analysis signals using the MATLAB program.

```
clc;
clear all;
close all;
t=-2:1:2;
x=[zeros(1,2) 1 ones(1,2)];
subplot (3,1,1);
stem(t,x);
grid;
x=[0 0 1 1 1];
disp(x);
title ('Impulse Response');
xlabel ('n');
ylabel ('Amplitude');
yn=x;
N=30;
xk=fft(yn,N);
```

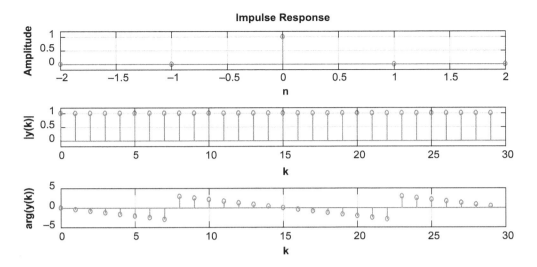

FIGURE 8.7 FFT of impulse sequence analysis signals in MATLAB.

```
magxk=abs(xk);
angxk=angle(xk);
k=0:N-1;
subplot(3,1,2);
stem(k,magxk);
grid;
xlabel('k');
ylabel('|y(k)|');
subplot(3,1,3);
stem(k,angxk);
disp(xk);
grid;
xlabel('k');
ylabel('arg(y(k))');
```

Figure 8.8 shows the FFT of unit step sequence analysis signals in MATLAB.

8.3 POWER SPECTRAL DENSITY USING SQUARE MAGNITUDE AND AUTOCORRELATION

To compute the power spectral density of the given signals using square magnitude and autocorrelation.

The stationary random processes do not have finite energy and later do not processes a Fourier Transform. Such signals have limited average power and hence are characterized by a power spectral density (PSD). The following program uses the property of PSD that is calculated from its autocorrelation function by taking the time average. In the square magnitude method, after taking the FFT by taking the absolute value one can get the PSD.

Program:

```
clc;
clear all;
close all;
f1=2000; %the frequency of first sequence in Hz
```

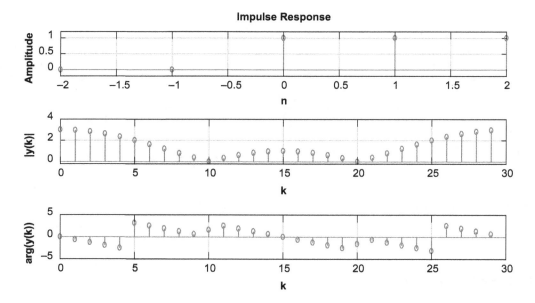

FIGURE 8.8 FFT of unit step sequence analysis signals in MATLAB.

```
f2=4000; %the frequency of the second sequence in Hz
fs=10000;%Enter the sampling frequency in Hz
t=0:1/fs:1;
x=2*sin(2*pi*f1*t)+3*sin(2*pi*f2*t)+rand(size(t));
px1=abs(fft(x).^2);
px2=abs(fft(xcorr(x),length(t)));
subplot(211)
plot(t*fs,10*log10(px1));%square magnitude
grid;
xlabel('Frequency (Hz)');
ylabel('Magnitude(dB)');
title('PSD');
subplot(212)
plot(t*fs,10*log10(px2));%autocorrelation
grid;
xlabel('Frequency (Hz)');
ylabel('Magnitude(dB)');
title('PSD');
```

Figure 8.9 shows an FFT power spectral density.

8.3.1 EQUIVALENCE OF FFT AND N-PHASE SEQUENCE COMPONENT TRANSFORMATION

We have seen that N-point DFT is given by the following expression

$$X(m) = \sum_{n=0}^{N-1} x_n e^{\frac{-j2\pi nm}{N}} \tag{8.1}$$

Let $a = e^{\frac{j2\pi}{N}}$ be the Nth root of the unit. Then the following relationships can be easily derived.

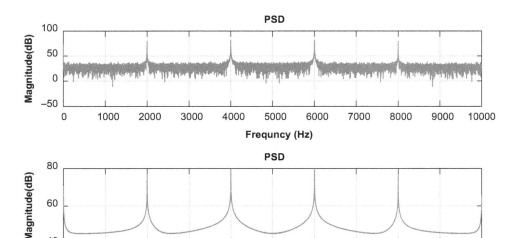

FIGURE 8.9   FFT Power spectral density.

1) $1 + a + a^2 + ----- + a^{N-1} = 0$ $\qquad\qquad$ (8.2)

Proof: using the geometric progression series formula

$$x + xr + xr^2 + ----- + xr^{N-1} = \frac{x(1-r^N)}{(1-r)} \qquad\qquad (8.3)$$

We $1 + a + a^2 + ----- + a^{N-1} = \dfrac{1(1-a^N)}{(1-a)} = \dfrac{(1-1)}{(1-a)} = 0$ because $a^N = 1$

2) From $a^m a^{N-m} = 1$ and $a^m (a^m)^* = 1$, we get

$$(a^m)^* = a^{N-m} \qquad\qquad (8.4)$$

Hence $a^* = a^{N-1}, (a^2)^* = a^{N-2}$ etc. $\qquad\qquad$ (8.5)

3) From the fact that $a^{-1}a = 1$ and $a^* a = 1$, we have

$$a^{-1} = a^* = a^{N-1} \text{ and } a^{-2} = (a^*)^2 = a^{N-2} \text{ etc.} \qquad\qquad (8.6)$$

Now using the a-operator, DFT transformation $m = 0, ----, N-1$ can be written as follows.

$$\begin{aligned} X(0) &= x_0 + x_1 + ----- + x_{N-1} \\ X(1) &= x_0 + x_1 a^{-1} + ---- + x_{N-1}(a^{-1})^{N-1} \\ \hline X(N-1) &= x_0 + x_1 a^{-(N-1)} + ----- + x_{N-1}(a^{-1})^{(N-1)^2} \end{aligned} \qquad (8.7)$$

Arranging it in a matrix format, we get the following

$$\begin{bmatrix} X(0) \\ X(1) \\ \vdots \\ X(N-1) \end{bmatrix} = \begin{bmatrix} 1 & 1 & - & - & 1 \\ 1 & a^{-1} & & & (a^{-1})^{N-1} \\ \vdots & \vdots & \ddots & & \vdots \\ 1 & (a^{-1})^{N-1} & - & - & (a^{-1})^{(N-1)^2} \end{bmatrix} \begin{bmatrix} x_0 \\ x_1 \\ \vdots \\ x_{N-1} \end{bmatrix} \qquad (8.8)$$

or stated more compactly

$$[X] = [P][x] \tag{8.9}$$

Where

$$P(i, j) = (a^{-1})^{(i-1)(j-1)} \tag{8.10}$$

The ith row of the matrix P indexes the ith frequency component, while the jth column of the $P-$ matrix indexes the jth sample. The matrix $P-$ enjoys an exceptional property viz. it's columns or rows that are orthogonal to each other. If p_i and p_j denote the ith and j^{th} column of the model P, then, it is easy to verify that

$$p_i^H p_j = 0 \quad (i \neq j) \tag{8.11}$$

$$p_i^H p_i = N \tag{8.12}$$

Where H indicates the Hermitian operator defined as

$$p_i^H = \left(p_i^T \right)^* = \left(p_i^* \right)^T \tag{8.13}$$

i.e., each column is first transposed into a row, and every element is then replaced by its complex conjugate.

For a real number, the complex conjugate is identical to the original number. Hence on real-valued vectors, Hermitian and transpose operators are the same. However, for complex-valued vectors, the two differ.

It is now easy to verify that

$$P^{-1} = \frac{1}{N} P^H \text{ and } P^H = \begin{bmatrix} 1 & 1 & 1 & - & 1 \\ 1 & a & a^2 & - & a^{N-1} \\ \vdots & \vdots & & & \vdots \\ \vdots & \vdots & & & \vdots \\ 1 & a^{(N-1)} & a^{2(N-1)} & - & a^{(N-1)^2} \end{bmatrix} \tag{8.14}$$

Thus,

$$[X] = [P][x] \tag{8.15}$$

$$[x] = \frac{1}{N} [P^H][X] \tag{8.16}$$

The invertibility of P, in essence, captures the transform property of DFT.

This effort is considered to be significantly high for real-time computing. However, with some ingenuity, researchers have shown that the task is achieved by approximate $(N/2) \log_2 N$ computations. This fast approach to computing all possible frequency transforms in the discrete domain is called the Fast Fourier Transform. For example, with $N=8$ brute force implementation of (1) requires 64 complex multiplications, which are reduced to 12 multiplications with FFT. As we would not have much use for FFT in this book, we will not pursue this topic any further. Rather, we will now establish an equivalence between two very well known transforms viz. multiple DFT or FFT and sequence component transformation.

We will first review the N-phase sequence transformation. Consider an N-phase (balanced or unbalanced) system ($N \geq 3$). Let the phasors in the phase domain (e.g., V_a, V_b, V_c for the three-phase system) be represented by $\bar{x}_1, \bar{x}_2, \ldots\ldots, \bar{x}_N$. Then, these phasors can be expressed as a linear combination of an N-set of balanced N-phase systems as follows.

0-sequence component

For n-phasor, the zero sequence system each of equal magnitude and angle

$$X_0 = \frac{\bar{x}_1 + \bar{x}_2 + ---- + \bar{x}_N}{N} \tag{8.17}$$

(+ve) 1-sequence component

$$X_1 = \frac{\bar{x}_1 + \bar{x}_2\, a + \bar{x}_3\, a^2 + ---- + \bar{x}_N\, a^{N-1}}{N} \tag{8.18}$$

For n-phasor, the positive sequence system. If $X_{1,1}$ is taken as a reference, then $X_{1,2}$ is equal in magnitude to $\bar{x}_1^{(a)}$ but by angle $\dfrac{2\pi}{N}$ i.e., $X_{1,2} = X_{1,1}\, a^{N-1}$

(−ve) 2-sequence component

$$X_2 = \frac{\bar{x}_1 + \bar{x}_2\, a^2 + \bar{x}_3\, a^4 + ---- + \bar{x}_N\, (a^2)^{N-1}}{N} \tag{8.19}$$

This system is obtained by relating lag in consecutive phases to the operator a^{-2} (or a^{N-2}).

$$X_{N-1} = \frac{\bar{x}_1 + \bar{x}_2\, (a^2)^{N-1} + \bar{x}_3\, (a^4)^{N-1} + ---- + \bar{x}_N\, a^{(N-1)^2}}{N} \tag{8.20}$$

Figure 8.10(a, b, and c) shows the symmetrical components system for a three-phase system.
Expressing these equations in a matrix format, we obtain the following equations:

$$\begin{bmatrix} X_0 \\ X_1 \\ | \\ | \\ X_{N-1} \end{bmatrix} = \frac{1}{N} \begin{bmatrix} 1 & 1 & 1 & - & 1 \\ 1 & a & a^2 & - & a^{N-1} \\ 1 & a^2 & a^4 & - & a^{2(N-1)} \\ | & | & & & | \\ 1 & a^{(N-1)} & a^{2(N-1)} & & a^{(N-1)^2} \end{bmatrix} \begin{bmatrix} \bar{x}_1 \\ \bar{x}_2 \\ | \\ | \\ \bar{x}_N \end{bmatrix} \tag{8.21}$$

Thus, writing compactly

$$[X] = [P^{-1}][\bar{x}]$$

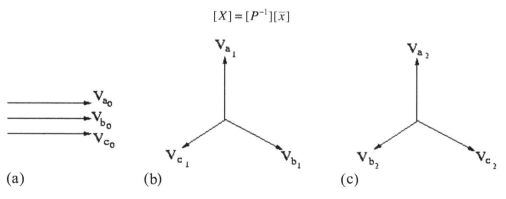

(a)   (b)   (c)

FIGURE 8.10   (a, b, and c) visualize the system for a three-phase system.

or

$$[\overline{x}] = [P][X] \tag{8.22}$$

Where

$$\overline{X} = \left[X_0, X_1, \ldots, X_{N-1}\right]^T \tag{8.23}$$

$$\overline{x} = \left[x_1, x_2, \ldots, x_N\right]^T \tag{8.24}$$

Thus, we conclude that FFT and sequence transformation (defined from the sequence domain to the phase domain) involve the same transformation matrix P. Hence, the two transforms are mathematically equivalent. In particular for $N = 3$, with a-phase as the reference phasor, we see the following equivalence relationships.

$$\begin{bmatrix} V_a \\ V_b \\ V_c \end{bmatrix} = \begin{bmatrix} 1 & 1 & 1 \\ 1 & a^2 & a \\ 1 & a & a^2 \end{bmatrix} \begin{bmatrix} V_0 \\ V_1 \\ V_2 \end{bmatrix} \tag{8.25}$$

and from DFT, we get

$$\begin{bmatrix} V(0) \\ V(1) \\ V(2) \end{bmatrix} = \begin{bmatrix} 1 & 1 & 1 \\ 1 & a^2 & a \\ 1 & a & a^2 \end{bmatrix} \begin{bmatrix} V_a \\ V_b \\ V_c \end{bmatrix} \tag{8.26}$$

This mathematical equivalence brings out an important concept viz. transformations and decompositions done via orthogonal matrices which can have multiple interpretations. However, there is one important difference between the two transformations. The samples $x_0, \ldots, x_{N-1}$ in signal processing are real numbers while corresponding phasors in sequence analysis are complex numbers. Thus, while there is a redundancy in information in the DFT domain which leads to DFT symmetry, there is no such redundancy in the sequence domain. Hence, in the sequence domain, we do not come across such a property.

PROBLEMS

8.1   Find the decimation-in-time FFT for the data sequence $x(0) = 0.5$, $x(1) = 0.25$, $x(2) = -0.25$, $x(3) = -0.5$, $x(4) = -0.5$, $x(5) = -0.25$, $x(6) = 0.25$, $x(7) = 0.5$.

8.2   Given eight samples values $x(0)$, $x(1), \ldots, x(7)$., draw a diagram to compute the DFT of the given sequence.
   a. Using the decimation-in-frequency FFT method.
   b. Using the decimation-in-time FFT method.

8.3   Compute the eight-point DFT of the sequence.

$$x(n) = \begin{cases} 1 & 0 \le n \le 3 \\ 0 & \text{otherwise} \end{cases}$$

   a. Using the decimation-in-frequency FFT method.
   b. Using the decimation-in-time FFT method.

8.4   Compute the eight-point DFT of the sequence

$$x(n) = \begin{cases} 1 & 0 \le n \le 7 \\ 0 & \text{otherwise} \end{cases}.$$

    a. Using the decimation-in-frequency FFT method.
    b. Using the decimation-in-time FFT method.

8.5   Compute the eight-point DFT of the sequence

$$x(n) = \begin{cases} a & 0 \le n \le 3 \\ 0 & \text{otherwise} \end{cases}.$$

Where a is a real number and constant.
    a. Using the decimation-in-frequency FFT method; and
    b. Using the decimation-in-time FFT method.

8.6   Compute the eight-point DFT of the sequence

$$x(n) = \begin{cases} a & 0 \le n \le 3 \\ 0 & \text{otherwise} \end{cases}$$

Where a is a real number and constant.
    a. Using the decimation-in-frequency FFT method; and
    b. Using the decimation-in-time FFT method.

8.7   Compute the eight-point inverse DFT of the sequence made up of DFT coefficients $X(k)$ as

$$X(0) = 4, X(1) = 1 - j\left(1 + \sqrt{2}\right), X(2) = 0, X(3) = 1 - j\left(1 - \sqrt{2}\right),$$

$$X(4) = 0, X(5) = 1 + j\left(1 - \sqrt{2}\right), X(6) = 0, X(7) = 1 + j\left(1 + \sqrt{2}\right).$$

    a. Using the decimation-in-frequency FFT method.
    b. Using the decimation-in-time FFT method.

8.8   Compute the eight-point inverse DFT of the sequence made up of DFT coefficients $X(k)$ as

$$X(0) = 0.25, X(1) = 0.125, X(2) = 0.125, X(3) = 0.25,$$

$$X(4) = 0.125, X(5) = 0.25, X(6) = 0.25, X(7) = 0.125$$

    a. Using the decimation-in-frequency FFT method.
    b. Using the decimation-in-time FFT method.

8.9   Compute the eight-point DFT of the sequence

$$x(0) = 1.1, x(1) = 1.1, x(2) = 1.1, x(3) = 1.1,$$

$$x(4) = 0, x(5) = 0, x(6) = 0, x(7) = 0.$$

    a. Using the decimation-in-frequency FFT method.
    b. Using the decimation-in-time FFT method.

8.10 Draw the eight-point DIT-FFT signal flow graph for the following sequence

$$X(n) = \left[0.5, 0.5, 0.5, 0.5, 0, 0, 0, 0\right]$$

8.11 Given a sequence $x(n)$. **For $0 \le n \le 3$, where $x(0) = 4, x(1) = 3, x(2) = 2$, and $x(3) = 1$,**
    a. Determine its DFT $X(k)$ using the decimation-in-frequency FFT method.
    b. Determine the number of complex multiplication in doing part (a).

8.12 Write a MATLAB program to find the autocorrelation of the following signal:
    $g(t) = e\text{-}2t\, u(t)$, Then, use the Wiener–Khintchine theorem to determine the energy spectral density of the signal.
    Use this program to display the autocorrelation function and the energy spectral density.

9 Z-Transform

The z-transform is useful for the manipulation of discrete data sequence and has acquired a new significance in the formulation and analysis of discrete-time systems. It is used extensively today in the area of applied mathematics digital signal processing, control theory, population science, economics. These discrete models are solved with difference equations in a manner that is analogous to solving continuous models with differential equations. The role played by the z-transform in the solution of difference equations corresponds to that represented by the Laplace transforms in the solution of differential equations. In this chapter the z-transform representation will present, also the Region of convergence (ROC), Properties of the z-Transform, and the Inverse z-Transform.

9.1 Z-TRANSFORM REPRESENTATION

In the previous chapters, we saw that the Discrete-Time Fourier Transform (DTFT) of sequence $x(n)$ is equal to the sum:

$$X(e^{j\omega}) = \sum_{n=-\infty}^{\infty} x(n) e^{-jn\omega} \tag{9.1}$$

However, for this series to converge, it is necessary that the signal is summable.

The z-transform is a generalization of the DTFT that allows one to deal with such sequences and is defined as follows:

Definition: The z-transform of a discrete-time signal $x(n)$ is characterized by:

$$X(z) = \sum_{n=-\infty}^{\infty} x(n) z^{-n} \tag{9.2}$$

In other words, the relation between $X(e^{j\omega})$ and $X(z)$ is:

$$X(e^{j\omega}) = X(z)\big|_{z=e^{j\omega}} \tag{9.3}$$

where $z = e^{j\omega}$ is a complex variable.

According to the fact that for most situations, the digital signal $x(n)$ is the causal sequence, that is, $x(n) = 0$ for $n < 0$, then $X(z)$ will be:

$$X(z) = \sum_{n=0}^{\infty} x(n) z^{-n} \tag{9.4}$$

Thus, the definition in the above equation is referred to as a one-sided z-transform or a unilateral transform.

Example 9.1

Prove that the convolution in the time domain implies convolution in the z-domain, i.e.,

$$Z(x_1(n) * x_2(n)) = Z(x_1(n)) Z(x_2(n)) = X_1(z)X_2(z)$$

SOLUTION

According to the definition of the convolution, we have

$$x(n) = x_1(n) * x_2(n) = \sum_{k=0}^{\infty} x_1(n-k)x_2(k)$$

Taking z-transform

$$X(n) = Z\left(x_1(n) * x_2(n)\right) = Z\left(\sum_{k=0}^{\infty} x_1(n-k)x_2(k)\right) = \sum_{n=0}^{\infty}\left(\sum_{k=0}^{\infty} x_1(n-k)x_2(k)\right)z^{-n}$$

It is written as

$$X(n) = \sum_{k=0}^{\infty} x_2(k)z^{-k} \cdot \sum_{n=0}^{\infty} x_1(n-k)z^{-(n-k)}$$

Let $n - k = m$, and then the second summation can be written as

$$X(n) = \sum_{k=0}^{\infty} x_2(k)z^{-k} \cdot \sum_{m=-k}^{\infty} x_1(m)z^{-m}$$

Using causality of both sequences, the second summation can be started from $m = 0$ instead of $m = -k$. Therefore, using the definition of z-transform, we have

$$X(n) = X_1(n) \cdot X_2(n)$$

Hence, proved.

9.2   REGION OF CONVERGENCE (ROC)

In Equation (9.2), all the values of z that make the summation exist (the sum converged) form a region of convergence in the z-transform domain. The region of convergence is defined based on the particular sequence $x(n)$ being applied. Note that we deal only with the unilateral z-transform, and hence when performing inverse z-transform (which we shall study later), we are restricted to the causal sequence.

Example 9.2

Given the sequence $x(n) = u(n)$, find the z-transform of $x(n)$.

SOLUTION

The z-transform is given by

$$X(z) = \sum_{n=0}^{\infty} u(n)z^{-n} = \sum_{n=0}^{\infty} \left(z^{-1}\right)^n$$

It is an infinite geometric series that converges to

$$X(z) = \frac{z}{z-1}$$

with a condition $\left|z^{-1}\right| < 1$.
The region of convergence for all values of z is given as $|z| > 1$

Example 9.3

Considering the exponential sequence $x(n) = a^n u(n)$, find the z-transform of the sequence $x(n)$.

SOLUTION

From the definition of the z-transform, it follows that

$$X(z) = \sum_{n=0}^{\infty} a^n u(n)z^{-n} = \sum_{n=0}^{\infty} \left(az^{-1}\right)^n$$

This geometric series which will converge for $\left|az^{-1}\right| < 1$

$$X(z) = \frac{z}{z-a}, \quad \text{for } |z| > |a|$$

The z-transform for common sequences is summarized in Table 9.1.

Example 9.4

Find the z-transform for each of the following sequences $x(n)$.

   a) $x(n) = 2.5 u(n)$

   b) $x(n) = 2\sin\left(\dfrac{\pi n}{4}\right)$

   c) $x(n) = (0.8)^n u(n)$

   d) $x(n) = (0.8)^n \sin\left(\dfrac{\pi n}{4}\right) u(n)$

   e) $x(n) = e^{-0.5n} \cos\left(\dfrac{\pi n}{4}\right) u(n)$

SOLUTION

   a) $x(n) = 2.5 u(n)$

      From Table 9.1

$$X(z) = Z[x(n)] = Z[2.5u(n)] = 2.5\frac{z}{z-1}$$

   b) $x(n) = 2\sin\left(\dfrac{\pi n}{4}\right)$

      From Table 9.1

$$X(z) = Z[x(n)] = Z\left[2\sin\left(\frac{\pi n}{4}\right)\right] = 2\frac{z\sin\left(\dfrac{\pi}{4}\right)}{z^2 - 2z\cos\left(\dfrac{\pi n}{4}\right) + 1}$$

   c) $x(n) = (0.8)^n u(n)$

$$X(z) = Z[x(n)] = Z[(0.8)^n u(n)] = \frac{z}{z-0.8}$$

TABLE 9.1
Table of z-Transform Pairs (for Causal Sequences)

No.	Signal $x(n), n \geq 0$	z-transform $Z(x(n)) = X(z)$	Region of Convergence
1	$x(n)$	$\displaystyle\sum_{n=0}^{\infty} x(n)z^{-n}$	
2	$\delta(n)$	1	Entire z-plane
3	$au(n)$	$\dfrac{az}{z-1}$	$|z| > 1$
4	$nu(n)$	$\dfrac{z}{(z-1)^2}$	$|z| > 1$
5	$n^2 u(n)$	$\dfrac{z(z+1)}{(z-1)^3}$	$|z| > 1$
6	$a^n u(n)$	$\dfrac{z}{z-a}$	$|z| > |a|$
7	$e^{-na} u(n)$	$\dfrac{z}{z-e^{-a}}$	$|z| > e^{-a}$
8	$na^n u(n)$	$\dfrac{az}{(z-a)^2}$	$|z| > |a|$
9	$\sin(an) u(n)$	$\dfrac{z\sin(a)}{z^2 - 2z\cos(a) + 1}$	$|z| > |1|$
10	$\cos(an) u(n)$	$\dfrac{z(z-\cos(a))}{z^2 - 2z\cos(a) + 1}$	$|z| > |1|$
11	$a^n \sin(bn) u(n)$	$\dfrac{[a\sin(b)]z}{z^2 - [2a\cos(b)]z + a^2}$	$|z| > |a|$
12	$a^n \cos(bn) u(n)$	$\dfrac{z[z - a\cos(b)]}{z^2 - [2a\cos(b)]z + a^2}$	$|z| > |a|$
13	$e^{-an} \sin(bn) u(n)$	$\dfrac{[e^{-a}\sin(b)]z}{z^2 - [2e^{-a}\cos(b)]z + e^{-2a}}$	$|z| > e^{-a}$
14	$e^{-an} \cos(bn) u(n)$	$\dfrac{z[z - e^{-a}\cos(b)]}{z^2 - [2e^{-a}\cos(b)]z + e^{-2a}}$	$|z| > e^{-a}$
15	$2|A||P|^n \cos(n\theta + \phi)u(n)$ where P and A are complex constants defined by $P = |P| \angle\theta,\ A = |A| \angle\phi$	$\dfrac{Az}{z-P} + \dfrac{A^*z}{z-P^*}$	

d) $x(n) = (0.8)^n \sin\left(\dfrac{\pi n}{4}\right) u(n)$

$$X(z) = Z\big[x(n)\big] = Z\left[(0.8)^n \sin\left(\dfrac{\pi n}{4}\right) u(n)\right]$$

$$= \dfrac{0.8\, z \sin\left(\dfrac{\pi}{4}\right)}{z^2 - 2 * 0.8\, z \cos\left(\dfrac{\pi n}{4}\right) + 0.8^2}$$

e) $x(n) = e^{-0.5n}\cos\left(\dfrac{\pi n}{4}\right) u(n)$

$$X(z) = Z\big[x(n)\big] = Z\left[e^{-0.5n}\cos\left(\dfrac{\pi n}{4}\right) u(n)\right]$$

$$= \dfrac{z\left[z - e^{-0.5}\cos\left(\dfrac{\pi}{4}\right)\right]}{z^2 - 2 * e^{-0.5} z \cos\left(\dfrac{\pi n}{4}\right) + (e^{-0.5})^2}$$

9.3 PROPERTIES OF THE Z-TRANSFORM

In this section, we study some important properties of the z-transform these properties are widely used in driving the z-transform functions of difference equations and solving the system output responses of linear digital systems with constant system coefficients.

For the following:

$$Z\{f[n]\} = \sum_{n=0}^{n=\infty} f[n]z^{-n} = F(z) \tag{9.5}$$

$$Z\{g_n\} = \sum_{n=0}^{n=\infty} g_n z^{-n} = G(z) \tag{9.6}$$

- Linearity

$$Z\{af_n + bg_n\} = aF(z) + bG(z). \tag{9.7}$$

and ROC is $R_f \cap R_g$,
which follows on from the definition of z-transform.

- Time shifting
  If we have $f[n] \Leftrightarrow F(z)$, then

$$f[n - n_0] \Leftrightarrow z^{-n_0} F(z) \tag{9.8}$$

The ROC of $Y(z)$ is the same as $F(z)$ except that there are possible pole additions or deletions at $z = 0$ or $z = \infty$.

Proof:

Let $y[n] = f[n - n_0]$, then

$$Y(z) = \sum_{n=-\infty}^{\infty} f\big[n - n_0\big] z^{-n} \tag{9.9}$$

Assume $k = n - n_0$, then $n = k + n_0$, substituting in the above equation we have:

$$Y(z) = \sum_{k=-\infty}^{\infty} f[k]z^{-k-n_0} = z^{-n_0}F[z] \qquad (9.10)$$

- Multiplication by an exponential sequence
  Let $y[n] = z_0^n f[n]$, then

$$Y(z) = X\left(\frac{z}{z_0}\right) \qquad (9.11)$$

The consequence is pole and zero locations that are scaled by z_0. If the ROC of $FX(z)$ is $r_R < |z| < r_L$, then the ROC of $Y(z)$ is

$$r_R < |z / z_0| < r_L, i.e., |z_0|r_R < |z| < |z_0|r_L$$

Proof:

$$Y(z) = \sum_{n=-\infty}^{\infty} z_0^n x[n]z^{-n} = \sum_{n=-\infty}^{\infty} x[n]\left(\frac{z}{z_0}\right)^{-n} = X\left(\frac{z}{z_0}\right) \qquad (9.12)$$

The consequence is pole and zero locations that are scaled by z_0. If the ROC of $X(z)$ is $rR<|z|< rL$, then the ROC of $Y(z)$ is

$$r_R < |z / z_0| < r_L, i.e., |z_0|r_R < |z| < |z_0|r_L$$

- Differentiation of $X(z)$
  If we have $f[n] \Leftrightarrow F(z)$, then

$$nf[n] \xleftrightarrow{\ z\ } -z\frac{dF(z)}{z} \quad \text{and} \quad \text{ROC} = R_f \qquad (9.13)$$

Proof:

$$F(z) = \sum_{n=-\infty}^{\infty} f[n]z^{-n} - z\frac{dF(z)}{dz}$$

$$= -z\sum_{n=-\infty}^{\infty} -n\, f[n]z^{-n-1} \qquad (9.14)$$

$$= \sum_{n=-\infty}^{\infty} -n\, f[n]z^{-n} - z\frac{dF(z)}{dz} \xleftrightarrow{\ z\ } nf[n]$$

- Conjugation of a Complex Sequence
  If we have $f[n] \Leftrightarrow F(z)$, then

$$f^*[n] \xleftrightarrow{\ z\ } F^*(z^*) \quad \text{and} \quad \text{ROC} = R_f \qquad (9.15)$$

Proof:
Let $y[n] = f^*[n]$, then

$$Y(z) = \sum_{n=-\infty}^{\infty} f^*[n]z^{-n} = \left(\sum_{n=-\infty}^{\infty} f[n][z^*]^{-n}\right)^* = F^*(z^*) \qquad (9.16)$$

- Time-reversal
  If we have $f[n] \Leftrightarrow F(z)$, then

$$f^*[-n] \xleftrightarrow{\;z\;} F^*\left(1/z^*\right) \qquad (9.17)$$

Let $y[n] = f^*[-n]$, then

$$Y(z) = \sum_{n=-\infty}^{\infty} f^*[-n]z^{-n} = \left(\sum_{n=-\infty}^{\infty} f[-n][z^*]^{-n}\right)^* = \left(\sum_{k=-\infty}^{\infty} f[k]\left(1/z^*\right)^{-k}\right)^* = F^*\left(1/z^*\right) \qquad (9.18)$$

If the ROC of $F(z)$ is $r_R < |z| < r_L$, then the ROC of $Y(z)$ is

$$r_R < \left|1/z^*\right| < r_L \text{ i.e., } \frac{1}{r_R} > |z| > \frac{1}{r_L}$$

When the time-reversal is without conjugation, it is easy to show

$$f[-n] \xleftrightarrow{\;z\;} F\left(1/z\right) \qquad (9.19)$$

and ROC is $\dfrac{1}{r_R} > |z| > \dfrac{1}{r_L}$

A comprehensive summary of the z-transform properties is shown in Table 9.2

Example 9.5

Find the z-transform of $x(n) = 3n + 2*3^n$.

SOLUTION

From the linearity property

$$Z[x(n)] = Z[3n + 2 \times 3^n] = 3Z[n] + 2Z[3^n]$$

$$Z\{n\} = \frac{z}{(z-1)^2} \text{ and } Z\{3^n\} = \frac{z}{(z-3)}$$

Therefore,

$$Z[x(n)] = \frac{3z}{(z-1)^2} + \frac{2z}{(z-3)}$$

Example 9.6

Find the z-transform of each of the following sequences:

(a) $x(n) = 2^n u(n) + 3\left(1/2\right)^n u(n)$

(b) $x(n) = \cos(5n)u(n)$.

SOLUTION

(a) Because $x(n)$ is a sum of two sequences of the form $a^n u(n)$, using the linearity property of the z-transform, and referring to Table 9.1, the z-transform pair

$$X(z) = \frac{1}{1 - 2z^{-1}} + \frac{3}{1 - \frac{1}{2}z^{-1}} = \frac{4 - \frac{13}{2}z^{-1}}{(1 - 2z)\left(1 - \frac{1}{2}z^{-1}\right)}$$

TABLE 9.2
Summary of z-Transform Properties

No.	Property	Sequence	z-Transform
1	Addition	$x_1(n) + x_2(n)$	$X_1(z) + X_2(z)$
2	Constant multiple	$cx(n)$	$c\,X(z)$
3	Linearity	$ax_1(n) + bx_2(n)$	$aX_1(z) + bX_2(z)$
4	Delay unit step	$u(n - m)$	$\dfrac{Z^{1-m}}{Z-1}$
5	Time delay shift	$x(n - m)\,U(n - m)$	$Z^{-m}\,X(z)$
6	Forward 1 tap	$x(n + 1)$	$Z(X(z) - x(0))$
7	Forward m tap	$x(n + 1)$	$Z^m\left(X(z) - \displaystyle\sum_{i=0}^{m-1} x_i Z^{-i}\right)$
8	Complex translation	$e^n\,x(n)$	$X(ze^{-1})$
9	Frequency scale	$b^n\,x(n)$	$X(zb^{-1})$
10	Differentiation	$n\,x(n)$	$-Z\,X'(z)$
11	Conjugation	$x^*(n)$	$X^*(z^*)$
12	Time-reversal	$x(-n)$	$X(z^{-1})$
13	Integration	$\dfrac{1}{n}x(n)$	$-\displaystyle\int \dfrac{X(z)}{Z}\,dz$
14	Discrete-time convolution	$x_1(n) * x_2(n)$	$X_1(z)X_2(z)$
15	Initial time	$x(0)$	$\displaystyle\lim_{n\to\infty} X(z)$
16	Final value	$\displaystyle\lim_{n\to\infty} x(n)$	$\displaystyle\lim_{n\to 1}(Z - 1)X(z)$

(b) For this sequence, we write

$$x(n) = \cos(5n)u(n) = \tfrac{1}{2}\left(e^{j5n} + e^{-j5n}\right)u(n)$$

Therefore, the z-transform is

$$X(z) = \frac{1}{2}\frac{1}{1 - e^{j5n}z^{-1}} + \frac{1}{2}\frac{1}{1 - e^{-j5n}z^{-1}}$$

With a region of convergence $|z| > 1$. Combining the two terms, we have

$$X(z) = \frac{1 - z^{-1}\cos 5}{1 - 2z^{-1}\cos 5 + z^{-2}}$$

Example 9.7

Find the z-transform of the sequence given by

$$x(n) = u(n) - (0.85)^n u(n)$$

SOLUTION

$$X(z) = Z\big[x(n)\big] = Z\big[u(n)\big] - Z\big[(0.85)^n u(n)\big]$$

$$= \frac{z}{z-1} - \frac{z}{z-0.85}$$

9.4   INVERSE Z-TRANSFORM

The z-transform of the sequence $x(n)$ and the inverse z-transform of the function $x(z)$ are defined as, respectively,

$$x(n) = \frac{1}{2\pi j} \oint X(z) z^{n-1} dz \qquad (9.20)$$

where the circular symbol on the integral sign denotes a closed counter in the complex plane.

The z-transform is a useful tool in linear systems analysis. However, just as important as techniques for finding the z-transform of a sequence are methods that may be used to invert the z-transform and recover the sequence $x(n)$ from $X(z)$. Three possible approaches are described below.

The inverse z-transform may be obtained by at least three methods:

1. Partial fraction expansion and look-up table
2. Power series expansion
3. Residue method

9.4.1   PARTIAL FRACTION EXPANSION AND A LOOK-UP TABLE

Now we are ready to deal with the inverse z-transform using the partial fraction, expansion, and look-up table. The general procedure is as follows:

1. Eliminate the negative powers of z for the z-transform function $X(z)$
2. Determine the rational function $X(z)\backslash z$ (assuming it is proper), and apply the partial fraction expanded function $X(z)\backslash z$ using the formula in Table 9.1.
3. Multiply the expanded function $X(z)\backslash z$ by z on both sides of the equation to obtain $X(z)$.
4. Apply the inverse z- transform using Table 9.1

The partial fraction format and the formula for calculating the constant are listed in Table 9.1.

For z-transforms that are rational functions of z,

$$X(z) = \frac{\sum_{k=0}^{q} b(k) z^{-1}}{\sum_{k=0}^{p} a(k) z^{-1}} = C \frac{\prod_{k=0}^{q} \left(1 - \beta_k z^{-1}\right)}{\prod_{k=1}^{p} \left(1 - \beta_k z^{-1}\right)} \qquad (9.21)$$

A simple and straightforward approach to find the inverse z-transform is to perform a partial fraction expansion of $X(z)$. Assuming that $p > q$ and that all of the roots in the denominator are simple, $\alpha_i \neq \alpha_k$ for $i \neq k$, $X(z)$ expands as follows:

$$X(z) = \sum_{k=1}^{p} \frac{A_k}{1 - \alpha_k z^{-1}} \qquad (9.22)$$

for some constants A_k for $k = 1,2, \ldots p$. The coefficients A_k may be found by multiplying both sides of Equation (9.3) by $(1 - \alpha_k z^{-1})$ and setting $z = \alpha_k$. The result is

$$A_k = \left[\left(1 - a_k z^{-1}\right) X(z)\right] \tag{9.23}$$

If $p \leq q$, the partial fraction expansion must include a polynomial in z^{-1} of order $(p - q)$. The coefficients of this polynomial found by long division (i.e., by dividing the numerator polynomial by the denominator). For multiple-order poles, the expansion must be modified. For example, if $X(z)$ has a second-order pole at $z = a_k$, the expansion will include two terms,

$$\frac{B_1}{1 - a_k z^{-1}} + \frac{B_2}{\left(1 - a_k z^{-1}\right)^2}$$

where B_1 and B_2 are given by

$$B_1 = a_k \left[\frac{d}{dz}\left(1 - a_k z^{-1}\right)^2 X(z)\right] \tag{9.24}$$

$$B_2 = \left[\left(1 - a_k z^{-1}\right)^2 X(z)\right] \tag{9.25}$$

Example 9.8

Find the inverse of the following z-transform

$$X(z) = \frac{1}{\left(1 - z^{-1}\right)\left(1 - 0.8 z^{-1}\right)}$$

SOLUTION

Multiplying the numerator and the denominator by z^2 we get

$$X(z) = \frac{1}{\left(1 - z^{-1}\right)\left(1 - 0.8 z^{-1}\right)} \times \frac{z^2}{z^2} = \frac{z^2}{(z - 1)(z - 0.8)}$$

Dividing both sides by z, we have

$$\frac{X(z)}{z} = \frac{z}{(z - 1)(z - 0.8)}$$

We notice that the right-hand side of the above equation is a proper rational polynomial of z. Also, we see that the denominator of the right-hand side has distinct poles, therefore, the right hand side in partial fraction form,

$$\frac{X(z)}{z} = \frac{A}{(z - 1)} + \frac{B}{(z - 0.8)}$$

To find out the unknown constants A and B, we use:

$$A = \left[(z - 1) \times \frac{X(z)}{z}\right]_{z=1} = \left[(z - 1) \times \frac{z}{(z - 1)(z - 0.8)}\right]_{z=1}$$

$$= \left[\frac{z}{(z - 0.8)}\right]_{z=1} = \frac{1}{(1 - 0.8)} = 5$$

$$B = \left[(z-0.8) \times \frac{X(z)}{z} \right]_{z=0.8} = \left[(z-0.8) \times \frac{z}{(z-1)(z-0.8)} \right]_{z=0.8}$$

$$= \left[\frac{z}{(z-1)} \right]_{z=0.8} = \frac{0.8}{(0.8-1)} = -4$$

Substituting the values, we have,

$$\frac{X(z)}{z} = \frac{5}{(z-1)} + \frac{(-4)}{(z-0.8)}$$

Or it can be written as (by multiplying both sides by z)

$$X(z) = \frac{5z}{(z-1)} - \frac{4}{(z-0.8)}$$

Taking the inverse z-transform of both sides and using Table 9.1, we have

$$x(n) = Z^{-1}(X(z)) = 5Z^{-1}\left(\frac{z}{(z-1)} \right) - 4Z^{-1}\left(\frac{z}{(z-0.8)} \right)$$

$$= 5\,u(n) - 4(0.8)^n\,u(n)$$

Example 9.10

Suppose that a sequence $x(n)$ has a z-transform

$$X(z) = \frac{4 - \frac{7}{4}z^{-1} + \frac{1}{4}z^{-2}}{1 - \frac{3}{4}z^{-1} + \frac{1}{8}z^{-2}}$$

SOLUTION

$$X(z) = \frac{4 - \frac{7}{4}z^{-1} + \frac{1}{4}z^{-2}}{1 - \frac{3}{4}z^{-1} + \frac{1}{8}z^{-2}} = \frac{4 - \frac{7}{4}z^{-1} + \frac{1}{4}z^{-2}}{\left(1 - \frac{1}{2}z^{-1}\right)\left(1 - \frac{1}{4}z^{-1}\right)}$$

with a region of convergence $|z| > \frac{1}{2}$. Because $p = q = 2$, and the two poles are simple, the partial fraction expansion has the form

$$X(z) = K_1 + \frac{K_2}{\left(1 - \frac{1}{2}z^{-1}\right)} + \frac{K_3}{\left(1 - \frac{1}{4}z^{-1}\right)}$$

The constant K_1 is found by long division, and equal to 2. K_2 and K_3 are equal to 3 and –1, respectively.

Thus, the complete partial fraction expansion becomes

$$X(z) = 2 + \frac{3}{\left(1 - \frac{1}{2}z^{-1}\right)} - \frac{1}{\left(1 - \frac{1}{4}z^{-1}\right)}$$

Finally, because the region of convergence is the exterior of the circle $|z| > 1$, $x(n)$ is the right-sided sequence

$$x(n) = 2\delta(n) + 3(0.5)^n u(n) - (0.25)^n u(n)$$

9.4.2 POWER SERIES

The z-transform is a power series expansion,

$$X(z) = \sum_{n=}^{\infty} x(n)z^{-n} = \cdots + x(-2)z^2 + x(-1)z^1$$

$$+ x(0) + x(1)z^{-1} + x(2)z^{-2} + \cdots$$

(9.26)

where the sequence values $x(n)$ are the coefficients of z^{-n} in the expansion; therefore, if we can find the power series expansion for $X(z)$, the sequence values $x(n)$ may be found by simply picking off the coefficients of z^{-n}.

Example 9.11

Consider the z-transform

$$X(z) = \log\left(\frac{Z+c}{Z}\right) \quad |Z| > |c|$$

SOLUTION

The power series expansion of this function is

$$X(z) = \log\left(\frac{Z+c}{Z}\right) = \log\left(1 + c Z^{-1}\right) \quad |Z| > |c|$$

$$= \sum_{n=1}^{\infty} \frac{1}{n}(-1)^{n+1} c^n Z^{-n} \quad n > 0$$

Therefore, the sequence $x(n)$ with this z-transform is

$$x(n) = \begin{cases} \dfrac{1}{n}(-1)^{n+1} c^n & n > 0 \\ 0 & \text{otherwise} \end{cases}$$

9.4.3 CONTOUR INTEGRATION

Another approach that may be used to find the inverse z-transform of $X(z)$ is to use contour integration. This procedure relies on Cauchy's integral theorem, which states that if C is a closed contour that encircles the origin in a counterclockwise direction,

$$\frac{1}{2\pi j} \oint_C z^{-k} \, dz = \begin{cases} 1 & k = 1 \\ 0 & k \neq 1 \end{cases}$$

(9.27)

with

$$X(z) = \sum_{n=-\infty}^{\infty} x(n) z^{-n}$$

(9.28)

Cauchy's integral theorem may be used to show that the coefficients $x(n)$ may be found from $X(z)$ as follows:

$$x(n) = \frac{1}{2\pi j} \oint_C X(z) z^{n-1} dz \tag{9.29}$$

where C is a closed contour within the region of convergence of $X(z)$ that encircles the origin in a counterclockwise direction. Contour integrals of this form may often be evaluated with the help of Cauchy's residue theorem,

$$x(n) = \frac{1}{2\pi j} \oint_C X(z) z^{n-1} dz = \sum \left(\text{residues of } X(z) z^{n-1} \text{ at the poles inside } C\right) \tag{9.30}$$

If $X(z)$ is a rational function of z with the first-order pole at $z = \alpha_k$,

$$x(n) = \frac{1}{2\pi j} \oint_C X(z) z^{n-1} dz = \sum \left(\text{residues of } X(z) z^{n-1} \text{ at the poles inside } C\right)$$

$$\text{Res}\left[X(z) z^{n-1} \text{ at } z = \alpha_k\right] = \left[\left(1 - z^{-1}\alpha_k\right) X(z) z^{n-1}\right] \tag{9.31}$$

Contour integration is particularly useful if only a few values of $x(n)$ are needed.

Example 9.12

Find the inverse of each of the following z-transforms:

i. $X(z) = 1 + 2\left(Z^2 + Z^{-2}\right) \quad 0 < |z| < \infty$

ii. $X(z) = \dfrac{1}{1 - 0.5Z^{-1}} + \dfrac{1}{1 - 0.2Z^{-1}} \quad 0.5 < |z|$

iii. $X(z) = \dfrac{1}{1 + 2Z^{-1} + Z^{-2}} \quad |z| > 2$

iv. $X(z) = \dfrac{1}{\left(1 - Z^{-1}\right)\left(1 - Z^{-2}\right)} \quad |z| > 1$

SOLUTION

i. Because $X(z)$ is a finite-order polynomial, $x(n)$ is a finite-length sequence. Therefore, $x(n)$ is the coefficient that multiplies z^{-1} in $X(z)$. Thus, $x(0) = 4$ and $x(2) = x(-2) = 3$.

ii. This z-transform is a sum of two first-order rational functions of z. Because the region of convergence of $X(z)$ is the exterior of a circle, $x(n)$ is a right-sided sequence. Using the z-transform pair for a right-sided exponential, we may invert $X(z)$ easily as follows:

$$x(n) = (0.5)^n u(n) + 3\left(\frac{1}{3}\right)^n u(n)$$

iii. Here we have a rational function of z with a denominator that is quadratic in z. Before we can find the inverse z-transform, we need to factor the denominator and perform a partial fraction expansion:

$$X(z) = \frac{1}{1 + 2Z^{-1} + Z^{-2}} = \frac{1}{\left(1 + 2Z^{-1}\right)\left(1 + Z^{-1}\right)}$$

$$X(z) = \frac{1}{\left(1 + 2Z^{-1}\right)} - \frac{1}{\left(1 + Z^{-1}\right)}$$

Because $x(n)$ is right-sided, the inverse z-transform is

$$x(n) = 2(-2)^n u(n) - (-1)^n u(n)$$

iv. One way to invert this z-transform is to perform a partial fraction expansion. With

$$X(z) = \frac{1}{\left(1 - Z^{-1}\right)\left(1 - Z^{-2}\right)}$$

The solution is given as follows:

$$x(n) = 0.25\left[1 + (-1)^n + 2(n + 1)\right]u(n)$$

Example 9.13

Find the inverse z-transform of the second-order system by

$$X(z) = \frac{1 + 0.25Z^{-1}}{\left(1 - 0.5Z^{-1}\right)^2} \quad |z| > 0.5$$

Here we have a second-order pole at $z = \frac{1}{2}$. The partial fraction expansion for $X(z)$ is

$$X(z) = \frac{K_1}{\left(1 - \frac{1}{2}z^{-1}\right)} + \frac{K_2}{\left(1 - \frac{1}{2}z^{-1}\right)^2}$$

The constants K_1 and K_2 are -0.5 and 1.5, respectively, so

$$X(z) = \frac{-0.5}{\left(1 - \frac{1}{2}z^{-1}\right)} + \frac{1.5}{\left(1 - \frac{1}{2}z^{-1}\right)^2}$$

and

$$x(n) = -(0.5)^{n+1}u(n) + 3(n + 1)(0.5)^{n+1}u(n)$$

Example 9.14

Find the inverse z-transform of $X(z) = \sin z$.

SOLUTION

To find the inverse z-transform of $X(z) = \sin z$, we expand $X(z)$ in a Taylor series of $z = 0$ as follows:
Because

$$X(z) = \sum_{n=-\infty}^{\infty} x(n)z^{-n}$$

So,

$$X(z) = X(z)\big|_{z=0} + z \frac{dX(z)}{dz}\bigg|_{z=0}$$

$$+ \frac{z^2}{2!} \frac{d^2 X(z)}{dz^2}\bigg|_{z=0} + \cdots + \frac{z^n}{n!} \frac{d^n X(z)}{dz^n}\bigg|_{z=0} + \cdots$$

$$= z - \frac{z^3}{3!} + \frac{z^5}{5!} - \cdots = \sum_{n=0}^{\infty} (-1)^n \frac{z^{2n+1}}{(2n+1)!}$$

Because

$$X(z) = \sum_{n=-\infty}^{\infty} x(n) z^{-n}$$

We may associate the coefficients in the Taylor series expansion with the sequence values $x(n)$. Thus, we have

$$x(n) = \sum_{n=0}^{\infty} (-1)^n \frac{z^{2n+1}}{(2n+1)!}$$

$$= (-1)^n \frac{1}{(2|n|+1)!} \quad n = -1, -3, -5, \ldots$$

Example 9.15

Evaluate the following integral:

$$\frac{1}{2\pi j} \oint_C \frac{1 + 2Z^{-1} - Z^{-2}}{\left(1 - 0.5Z^{-1}\right)\left(1 - \frac{2}{3}Z^{-1}\right)} z^3 dz$$

where the contour of integration C is the unit circle.

SOLUTION

Recall that for a sequence $x(n)$ that has a z-transform $X(z)$, and the sequence may be recovered using contour integration as follows:

$$x(n) = \frac{1}{2\pi j} \oint_C X(z) z^{n-1} dz$$

Therefore, the integral that is to be evaluated corresponds to the value of the sequence $x(n)$ at $n = 4$ that has a z-transform

$$X(z) = \frac{1 + 2Z^{-1} - Z^{-2}}{\left(1 - 0.5Z^{-1}\right)\left(1 - \frac{2}{3}Z^{-1}\right)}$$

Thus, we may find $x(n)$ using a partial fraction expansion of $X(z)$ and then evaluate the sequence at $n = 4$. With this approach, however, we are finding the values of $x(n)$ for all of n. Alternatively, we could perform long division and divide the numerator of $X(z)$ by the denominator. The coefficient multiplying z^{-4} would then be the value of $x(n)$ at $n = 4$, and the value of the integral. However, because we are only interested in the value of the sequence at $n = 4$, the easiest approach is to evaluate the integral directly using Cauchy's integral theorem. The value of the integral is equal to the sum of the residues of the poles of $X(z)z^3$ inside the unit circle. Because

$$X(z)z^3 = \frac{1 + 2Z^{-1} - Z^{-2}}{\left(1 - 0.5Z^{-1}\right)\left(1 - \frac{2}{3}Z^{-1}\right)} z^3$$

has poles at $z = 1/2$ and $z = 2/3$,

$$\text{REs}\left[X(z)z^3\right]\Big|_{z=0.5} = -\frac{3}{16}$$

and

$$\text{REs}\left[X(z)z^3\right]\Big|_{z=2/3} = \frac{112}{81}$$

Therefore, we have

$$\frac{1}{2\pi j}\oint_C X(z)\,z^{n-1}dz = -\frac{3}{16}+\frac{112}{81} = 1.195$$

Example 9.16

Find the *z*-transform for the following sequences:

    a. $x(n) = 15\,u(n)$

    b. $x(n) = 10\sin(0.25\pi n)u(n)$

    c. $x(n) = (0.5)^n u(n)$

    d. $x(n) = (0.5)^n \sin(0.25\pi n)u(n)$

    e. $x(n) = e^{-0.1n}\cos(0.25\pi n)u(n)$

SOLUTION

(a) From Table 9.1, we get

$$X(z) = Z\big(x(n)\big) = Z\big(15u(n)\big) = \frac{15}{z-1}$$

(b) From Table 9.1, we obtain

$$X(z) = Z\big(x(n)\big) = Z\big(10\sin(0.25\pi n)u(n)\big)$$

$$= \frac{10\sin(0.25\pi)z}{z^2-2z\cos(0.25\pi n)+1} = \frac{7.07\,z}{z^2-1.414z+1}$$

(c) From Table 9.1, we get

$$X(z) = Z\big(x(n)\big) = Z\big((0.5)^n u(n)\big) = \frac{z}{z-0.5}$$

(d) From Table 9.1, we get

$$X(z) = Z\big(x(n)\big) = Z\big((0.5)^n \sin(0.25\pi n)u(n)\big)$$

$$= \frac{0.5\times\sin(0.25\pi)z}{z^2-2\times0.5\times z\cos(0.25\pi n)+(0.5)^2}$$

$$= \frac{0.3536\,z}{z^2-1.4142z+0.25}$$

(e) From Table 9.1, we get

$$X(z) = Z\big(x(n)\big) = Z\big(e^{-0.1n}\cos(0.25\pi n)u(n)\big)$$

$$= \frac{z\big(z - e^{-0.1}\cos(0.25\pi)\big)}{z^2 - 2z\,e^{-0.1}\cos(0.25\pi) + \big(e^{-0.1}\big)^2}$$

$$= \frac{z(z - 0.6397)}{z^2 - 1.279\,z + 0.8187}$$

Example 9.17

Find the inverse of the following z-transform,

$$X(z) = \frac{1}{\big(1 - z^{-1}\big)\big(1 - 0.5z^{-1}\big)}$$

Eliminating the negative power of z by multiplying by z^2 and determine by

$$X(z) = \frac{z^2}{z^2\big(1 - z^{-1}\big)\big(1 - 0.5z^{-1}\big)}$$

Dividing both sides by z leads to

$$\frac{X(z)}{z} = \frac{z}{(z - 1)(z - 0.5)}$$

$$= \frac{A}{(z - 1)} + \frac{B}{(z - 0.5)}$$

$$A = \lim_{z=1}\frac{z}{(z - 0.5)} = 2$$

$$B = \lim_{z=0.5}\frac{z}{(z - 1)} = 1$$

$$x(n) = Z^{-1}\big(X(z)\big) = Z^{-1}\left(\frac{z}{(z - 1)}\right) + Z^{-1}\left(\frac{z}{(z - 0.5)}\right) = u(n) + (0.5)^n u(n)$$

Example 9.18

Find the inverse of the following z-transform

i. $X(z) = 1 + \dfrac{z}{(z - 1)} + \dfrac{z}{(z - 0.8)}$

ii. $X(z) = \dfrac{2z}{(z - 1)^2} + \dfrac{5z}{(z - 0.8)^2}$

iii. $X(z) = \dfrac{z^{-6}}{z + 1} + \dfrac{z^{-5}}{z - 0.7} + z^{-4}$

SOLUTION

Dividing both sides by z, we have

i. $X(z) = 1 + \dfrac{z}{(z - 1)} + \dfrac{z}{(z - 0.8)}$

$$x(n) = Z^{-1}(X(z))$$

$$= Z^{-1}(1) + Z^{-1}\left(\frac{z}{(z-1)}\right) + Z^{-1}\left(\frac{z}{(z-0.8)}\right)$$

$$= \delta(n) + u(n) + (0.8)^n u(n)$$

ii. $X(z) = \dfrac{2z}{(z-1)^2} + \dfrac{5z}{(z-0.8)^2}$

$$x(n) = Z^{-1}(X(z)) = Z^{-1}\left(\frac{2z}{(z-1)^2}\right) + Z^{-1}\left(\frac{5z}{(z-0.8)^2}\right)$$

$$= 2n\,u(n) + 5n(0.8)^n\,u(n)$$

$$= 2r(n) + 5(0.8)^n\,r(n)$$

iii. $X(z) = \dfrac{z^{-6}}{z+1} + \dfrac{z^{-5}}{z-0.7} + z^{-4}$

$$x(n) = Z^{-1}(X(z)) = Z^{-1}\left(\frac{z^{-6}}{z+1}\right) + Z^{-1}\left(\frac{z^{-5}}{z-0.7}\right) + Z^{-1}(z^{-4})$$

$$x(n) = Z^{-1}\left(z^{-7}\frac{z}{z+1}\right) + Z^{-1}\left(z^{-6}\frac{z}{z-0.7}\right) + Z^{-1}(z^{-4})$$

$$x(n) = u(n-7) + (0.7)^n u(n-6) + \delta(n-4)$$

Example 9.19

Determine the convolution of the following two sequences, using z-transform,

$$x_1(n) = 3\delta(n) + 2\delta(n-1)$$

$$x_2(n) = 2\delta(n) - \delta(n-1)$$

SOLUTION

Taking the z-transform of the two sequences, we have

$$X_1(z) = Z(3\delta(n) + 2\delta(n-1)) = 3 + 2z^{-1}$$

$$X_2(z) = Z(2\delta(n) - \delta(n-1)) = 2 - z^{-1}$$

Using the z-transform property for convolution of two sequences, we have

$$X(z) = X_1(z)\,X_2(z) = (3 + 2z^{-1})(2 - z^{-1})$$

Therefore,

$$X(z) = 6 + z^{-1} - 2z^{-2}$$

Taking inverse z-transform of both sides (and using shift theorem), we have

$$x(n) = Z^{-1}(X(z)) = 6Z^{-1}(1) + Z^{-1}(z^{-1}) - 2Z^{-1}(z^{-2})$$

$$= 6\delta(n) + \delta(n-1) - 2\delta(n-2)$$

Example 9.20

Given a transfer function depicting a DSP system

$$H(z) = \frac{z-1}{z+0.5}$$

Determine

(a) The impulse response $h(n)$.
(b) The step response $s(n)$.
(c) The system response $y(n)$, if the input is given as $x(n) = (0.5)^n u(n)$.

SOLUTION

Part (a): In this case $x(n) = \delta(n)$, thus $X(z) = 1$. As

$$H(z) = \frac{Y(z)}{X(z)}$$

Therefore, in this case, the z-transform of the output is equal to the transfer function:

$$H(z) = Y(z)$$

By taking the inverse z-transform of the transfer function, we can find out the unit-impulse response $h(n)$, of the system. The transfer function is written as

$$\frac{H(z)}{z} = \frac{z-1}{z(z+0.5)}$$

This can further be written in the form of partial fractions as

$$\frac{H(z)}{z} = \frac{A}{z} + \frac{B}{(z+0.5)}$$

where

$$A = \frac{z-1}{(z+0.5)}\bigg|_{z=0} = \frac{0-1}{(0+0.5)} = -2$$

$$B = \frac{z-1}{z}\bigg|_{z=0.5} = \frac{0.5-1}{0.5} = -1$$

Thus, we have

$$\frac{H(z)}{z} = \frac{-2}{z} + \frac{-1}{(z+0.5)}$$

or

$$H(z) = -2 - \frac{z}{(z-0.5)}$$

Taking inverse z-transform of both sides (and using Table 5.1), we get

$$h(n) = -2\delta(n) - (0.5)^n u(n)$$

which is the required impulse response of the system.

Part (b): In this case $x(n) = u(n)$, thus $X(z) = \frac{z}{z-1}$. As

$$H(z) = \frac{Y(z)}{X(z)}$$

Therefore, in this case,

$$Y(z) = H(z)X(z) = \frac{(z-1)}{(z+0.5)} \frac{z}{(z-1)}$$

It is written as

$$\frac{Y(z)}{z} = \frac{z-1}{(z+0.5)(z-1)} = \frac{1}{(z+0.5)}$$

Taking inverse z-transform of both sides, we get

$$y(n) = (-0.5)^n u(n)$$

which is the required step response of the system.

Part (c): In this case $x(n) = (0.25)^n u(n)$, $X(z) = \dfrac{z}{z-0.25}$

$$H(z) = \frac{Y(z)}{X(z)}$$

Therefore, in this case,

$$Y(z) = H(z)X(z) = \frac{(z-1)}{(z+0.5)} \cdot \frac{z}{(z-0.25)}$$

It is written as

$$\frac{Y(z)}{z} = \frac{z-1}{(z+0.5)(z-0.25)}$$

Further, it can be written in the form of partial fractions as

$$\frac{Y(z)}{z} = \frac{A}{(z+0.5)} + \frac{B}{(z-0.25)}$$

where

$$A = \frac{z-1}{(z-0.25)}\bigg|_{z=-0.5} = \frac{-0.5-1}{(-0.5-0.25)} = 2$$

$$B = \frac{z-1}{(z+0.5)}\bigg|_{z=0.25} = \frac{0.25-1}{0.25+0.5} = -1$$

Thus, we have

$$\frac{Y(z)}{z} = \frac{2}{(z+0.5)} + \frac{-1}{(z-0.25)}$$

or

$$Y(z) = \frac{2z}{(z+0.5)} - \frac{z}{(z-0.25)}$$

Taking inverse z-transform of both sides, we get

$$y(n) = 2(-0.5)^n u(n) - (0.25)^n u(n)$$

PROBLEMS

9.1 Given the sequence $x(n) = u(n-1)$, find the z-transform of $x(n)$.

9.2 Given the sequence $x(n) = u(n+1)$, find the z-transform of $x(n)$.

9.3 Considering the exponential sequence $x(n) = a^n u(n-1)$, find the z-transform of the sequence $x(n)$.

9.4 Considering the exponential sequence $x(n) = a^n u(n+1)$, find the z-transform of the sequence $x(n)$.

9.5 Find the z-transform for each of the following sequences $x(n)$.

a) $x(n) = 2.5 u(n-1)$

b) $x(n) = 102 \sin\left(\dfrac{\pi n}{4}\right)$

c) $x(n) = (0.8)^n u(n-1)$

d) $x(n) = (0.8)^{n-1} \sin\left(\dfrac{\pi n}{4}\right) u(n-1)$

e) $(n) = e^{-0.5n} \cos\left(\dfrac{\pi n}{4}\right) u(n-1)$

9.6 Find the z-transform of $x(n) = 3n + 2 * 3^{n-1}$.

9.7 Find the z-transform of each of the following sequences:

a) $x(n) = 2^n u(n-1) + 3(1/2)^n u(n-1)$

b) $x(n) = \cos(5n) u(n-1)$

9.8 Find the z-transform of the sequence given by

$$x(n) = u(n+1) - (0.85)^n u(n-1)$$

9.9 Find the z-transform of the sequence given by

$$x_1(n) = \delta(n) - \delta(n-1) + \delta(n-2)$$

$$x_2(n) = 1.5\,\delta(n) + 2\,\delta(n-1)$$

Also, find the convolution $x(n) = x_1(n) * x_2(n)$

9.10 Find the inverse of the following z-transform

$$X(z) = \frac{10}{\left(1 - 0.3z^{-1}\right)\left(1 - 0.7z^{-1}\right)}$$

9.11 Find the z-transform

$$X(z) = \log\left(\frac{Z+c}{Z}\right) \quad |Z| > |c|$$

9.12 Find the inverse of each of the following z-transforms:

i. $X(z) = 1 + 2\left(Z^3 + Z^{-3}\right) \quad 0 < |z| < \infty$

ii. $X(z) = \dfrac{10}{1 - 0.5Z^{-1}} - \dfrac{10}{1 - 0.2Z^{-1}} \quad 0.5 < |z|$

iii. $X(z) = \dfrac{10Z}{1 + 2Z^{-1} + Z^{-2}}$ $|z| > 2$

iv. $X(z) = \dfrac{1}{Z\left(1 - Z^{-1}\right)\left(1 - Z^{-2}\right)}$ $|z| > 1$

9.13   Find the inverse z-transform of $X(z) = \cos z$.

9.14   Given a transfer function depicting a DSP system

$$H(z) = \frac{z - 1}{z(z + 0.5)}$$

Determine

(a) The impulse response $h(n)$.
(b) The step response $s(n)$.
(c) The system response $y(n)$, if the input is given as $x(n) = (0.8)^n u(n)$.

9.15   Given a transfer function depicting a DSP system

$$H(z) = \frac{z(z - 1)}{(z + 0.5)}$$

Determine

(a) The impulse response $h(n)$.
(b) The step response $s(n)$.
(c) The system response $y(n)$, if the input is given as $x(n) = (0.75)^n u(n)$.

9.16   Find the z-transform for each of the following sequences (from the definition of the z-transform),

a. $x(n) = 4 u(n)$

b. $x(n) = (0.7)^n u(n)$

c. $x(n) = e^{-2n} u(n)$

d. $x(n) = 2(0.8)^n \cos(0.2\pi n) u(n)$

e. $x(n) = 2.5 e^{-3n} \sin(0.2\pi n) u(n)$

9.17   Using the properties of the z-transform, find the z-transform for each of the following sequences:

a. $x(n) = u(n - 1) + (0.5)^n u(n)$

b. $x(n) = e^{-3(n-5)} \cos\left(0.1\pi(n - 5)\right) u(n - 5)$

where $u(n - 5) = 1$ for $n \geq 5$ and $u(n - 5) = 0$ for $n < 5$.

10 Z-Transform Applications in DSP

Z- transform is used in many applications of mathematics and signal processing. The applications of z transform are Analyze the discrete linear system, finding frequency response Analysis of discrete signal, Helps in system design and analysis and also checks the systems stability and analysis of digital filters. This chapter will focus on some these applications

10.1 EVALUATION OF LTI SYSTEM RESPONSE USING Z-TRANSFORM

Figure 10.1 shows an LTI system where $h(n)$ is the impulse response of the system, and $H(z)$ is the transfer function, the input signal $x(n)$ and the output response is $y(n)$, the transfer function in z-domain is given as

$$H(z) = H\left(e^{j\omega}\right) = \frac{Y(z)}{X(z)} \tag{10.1}$$

10.2 DIGITAL SYSTEM IMPLEMENTATION FROM ITS FUNCTION

The z-transform is a linear transformation, the system implementation produce is similar to that in the time domain. The most convenient form for system synthesis is the z-transform of the general difference equation given by

$$Y(z) = \sum_{k=1}^{M} a_k z^{-k} Y(z) + \sum_{k=q}^{b} b_k z^{-k} X(z) \tag{10.2}$$

1 Gain: Figure 10.2 shows a gain block, where k is the value of gain.
2 Delay: Figure 10.3 shows a delay block.
3 Addition: Figure 10.4 shows an addition block used to add two or more signals.

Example 10.1

Find the impulse response and the transfer function of the following system as shown in Figure 10.5.

SOLUTION

$$y(n) = x(n) + K\,y(n-1)$$

$$Y(z) = X(z) + K\,Y(z)z^{-1}$$

For impulse response $x(n) = \delta(n)$
  $X(z) = 1$

$$Y(z) = 1 + K\,Y(z)z^{-1}$$

$$Y(z) = \frac{1}{1 - K\,z^{-1}}$$

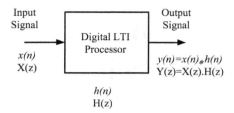

FIGURE 10.1   LTI processor system.

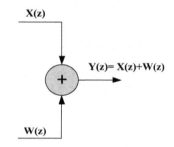

FIGURE 10.2   Gain block.

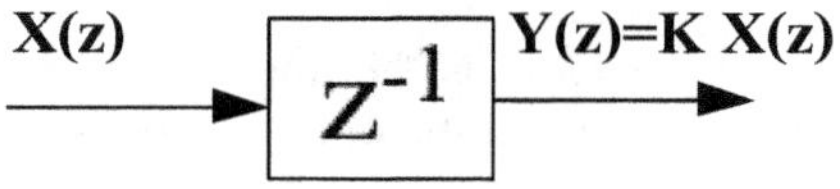

FIGURE 10.3   Delay block.

FIGURE 10.4   Addition block.

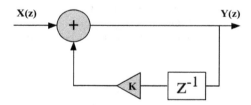

FIGURE 10.5   System for Example 10.1.

Example 10.2

Implement a second-order recursive filter for the sequence

$$y(n) = 2r\cos\omega_o\, y(n-1) - r^2\, y(n-2) + x(n) - r\cos\omega_o\, x(n-1)$$

SOLUTION

Take z-transform for both sides to get

$$Y(z) = 2r\cos\omega_o Y(z)z^{-1} - r^2\, Y(z)z^{-2} + X(z) - r\cos\omega_o\, X(z)z^{-1}$$

So the system implement as shown in Figure 10.6.

Example 10.3

Given a second-order transfer function

$$H(z) = \frac{2.5\left(1-z^{-1}\right)}{\left(1+1.3\,z^{-1}+0.36\,z^{-2}\right)}$$

Perform the filter realizations and write the difference equations using the cascade form realizations via the first-order sections.

SOLUTION

For the cascade realization, the transfer function is written in the product form. The given transfer function is

$$H(z) = \frac{2.5\left(1-z^{-2}\right)}{1+1.3\,z^{-1}+0.36\,z^{-2}}$$

The numerator polynomial is factorized as

$$B(z) = 2.5\left(1-z^{-2}\right) = 2.5\left(1-z^{-1}\right)\left(1+z^{-1}\right)$$

The denominator polynomial is factorized as

$$A(z) = 1+1.3\,z^{-1}+0.36\,z^{-2} = 1+0.4\,z^{-1}+0.9\,z^{-1}+0.36\,z^{-2}$$

$$= 1\left(1+0.4\,z^{-1}\right)+0.9\,z^{-1}\left(1+0.4\,z^{-1}\right) = \left(1+0.4\,z^{-1}\right)\left(1+0.9\,z^{-1}\right)$$

Therefore, the transfer function is written as

$$H(z) = \frac{2.5\left(1-z^{-1}\right)\left(1+z^{-1}\right)}{\left(1+0.4\,z^{-1}\right)\left(1+0.9\,z^{-1}\right)}$$

Or

$$H(z) = \left(\frac{2.5-2.5\,z^{-1}}{1+0.4\,z^{-1}}\right)\left(\frac{1+z^{-1}}{1+0.9\,z^{-1}}\right) = H_1(z)\cdot H_2(z)$$

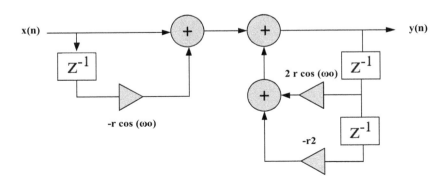

FIGURE 10.6 System for Example 10.2.

Thus, in this case

$$H_1(z) = \frac{2.5 - 2.5\,z^{-1}}{1 + 0.4\,z^{-1}}$$

$$H_2(z) = \frac{1 + z^{-1}}{1 + 0.9\,z^{-1}}$$

Each one of $H_1(z)$ and $H_2(z)$ can be realized in direct-form I or direct-form II. Overall, we get the cascaded realization with two sections. It should be noted that there could be other forms for $H_1(z)$ and $H_2(z)$, for example, we could have taken $H_1(z) = \dfrac{1 + z^{-1}}{1 + 0.4\,z^{-1}}$, $H_2(z) = \dfrac{2.5 - 2.5\,z^{-1}}{1 + 0.9\,z^{-1}}$, to yield the same $H(z)$. Using the former $H_1(z)$ and $H_2(z)$, and using direct-form II realizations for the two cascaded sections as shown in Figure 10.7, we get the following difference equations:

Section 1: $\left(H_1(z) = \dfrac{2.5 - 2.5\,z^{-1}}{1 + 0.4\,z^{-1}} \right)$

$$w_1(n) = x(n) - 0.4w(n-1)$$

$$y_1(n) = 2.5w_1(n) - 2.5w_1(n-1)$$

Section 2: $\left(H_2(z) = \dfrac{1 + z^{-1}}{1 + 0.9\,z^{-1}} \right)$

$$w_2(n) = y_1(n) - 0.9w_2(n-1)$$

$$y(n) = w_2(n) + w_2(n-1)$$

Example 10.4

A relaxed (zero initial conditions) DSP system is described by the difference equation

$$y(n) + 0.1y(n-1) - 0.2y(n-2) = x(n) + x(n-1)$$

Determine the impulse response $y(n)$ due to the impulse sequence $x(n) = \delta(n)$.

SOLUTION

Taking the z-transform of both sides of the given equation, we get

$$Z\big(y(n)\big) + 0.1Z\big(y(n-1)\big) - 0.2Z\big(y(n-2)\big) = Z\big(x(n)\big) + Z\big(x(n-1)\big) \qquad (10.1)$$

We have

$$Z(y(n)) = Y(z)$$

$$Z(x(n)) = X(z)$$

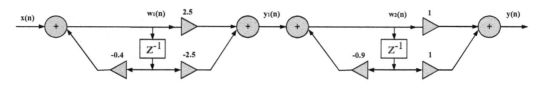

FIGURE 10.7   System for Example 10.3.

Using shift theorem, we have

$$Z\big(x(n-1)\big) = z^{-1}X(z)$$

Also, we can apply to sift the theorem for y in a case of zero initial conditions, i.e.,

$$Z\big(y(n-1)\big) = z^{-1}Y(z)$$

$$Z\big(y(n-2)\big) = z^{-2}Y(z)$$

Putting these values in Equation (10.1), we have

$$Y(z) + 0.1z^{-1}Y(z) - 0.2\,z^{-2}Y(z) = X(z) + z^{-1}X(z)$$

$$\Rightarrow Y(z)\big(1 + 0.1z^{-1} - 0.2\,z^{-2}\big) = X(z)\big(1 + z^{-1}\big)$$

As $x(n) = \delta(n)$ therefore (from Table 9.1), $X(z) = 1$. The above equation can now be written as

$$Y(z) = \frac{\big(1 + z^{-1}\big)}{\big(1 + 0.1z^{-1} - 0.2z^{-2}\big)}$$

Multiplying both the numerator and the denominator with z^2, we get

$$Y(z) = \frac{z(z+1)}{\big(z^2 + 0.1z - 0.2\big)}$$

The denominator is factorized as

$$Y(z) = \frac{z(z+1)}{\big(z^2 + 0.5\,z - 0.4\,z - 0.2\big)} = \frac{z(z+1)}{\big(z(z+0.5) - 0.4(z+0.5)\big)} = \frac{z(z+1)}{(z+0.5)(z-0.4)}$$

(10.2)

$$\Rightarrow \frac{Y(z)}{z} = \frac{(z+1)}{(z+0.5)(z-0.4)}$$

The right-hand side of the above equation is a proper rational polynomial, with the denominator polynomial having distinct poles. Therefore, it can be written into partial fractions as

$$\frac{Y(z)}{z} = \frac{A}{(z+0.5)} + \frac{B}{(z-0.4)}$$

(10.3)

To find out the unknown constants A and B, we use:

$$A = \left[(z+0.5) \times \frac{X(z)}{z}\right]_{z=-0.5}$$

$$= \left[(z+0.5) \times \frac{(z+1)}{(z+0.5)(z-0.4)}\right]_{z=-0.5}$$

$$= \left[\frac{(z+1)}{(z-0.4)}\right]_{z=-0.5} = \frac{(-0.5+1)}{(-0.5-0.4)} = \frac{0.5}{-0.9} = -0.5556$$

$$B = \left[(z-0.4) \times \frac{X(z)}{z}\right]_{z=0.4} = \left[(z-0.4) \times \frac{(z+1)}{(z+0.5)(z-0.4)}\right]_{z=0.4}$$

$$= \left[\frac{(z+1)}{(z+0.5)}\right]_{z=0.4} = \frac{(0.4+1)}{(0.4+0.5)} = \frac{1.4}{0.9} = 1.5556$$

Equation (10.3) becomes:

$$\frac{Y(z)}{z} = \frac{-0.5556}{(z+0.5)} + \frac{1.5556}{(z-0.4)}$$

$$Y(z) = \frac{-0.5556\,z}{(z+0.5)} + \frac{1.5556\,z}{(z-0.4)}$$

Taking an inverse z-transform of both sides

$$y(n) = Z^{-1}(Y(z)) = Z^{-1}\left(\frac{-0.5556z}{(z+0.5)}\right) + Z^{-1}\left(\frac{1.5556z}{(z-0.4)}\right)$$

$$= (-0.5556)\,Z^{-1}\left(\frac{z}{(z-(-0.5))}\right) + (1.5556)\,Z^{-1}\left(\frac{z}{(z-0.4)}\right)$$

$$= (-0.5556)(-0.5)^n\,u(n) + (1.5556)(0.4)^n\,u(n)$$

Thus, the output signal is

$$y(n) = (-0.5556)(-0.5)^n\,u(n) + (1.5556)(0.4)^n\,u(n)$$

Example 10.5

A relaxed (zero initial conditions) DSP system is described by the difference equation

$$y(n) = 0.4\,y(n-1) + 0.32\,y(n-2) + x(n) + 0.1x(n-1)$$

Determine the impulse response $y(n)$ due to the impulse sequence $x(n) = \delta(n)$.

SOLUTION

$$y(n) - 0.4\,y(n-1) - 0.32\,y(n-2) + x(n) + 0.1x(n-1)$$

Taking a z-transform of both sides of the given equation, we get

$$Z\big(y(n)\big) - 0.4Z\big(y(n-1)\big) - 0.32Z\big(y(n-2)\big) = Z\big(x(n)\big) + 0.1Z\big(x(n-1)\big)$$

$$Y(z) - 0.4\,z^{-1}Y(z) - 0.32\,z^{-2}Y(z) = X(z) + 0.1z^{-1}X(z)$$

$$\Rightarrow Y(z)\big(1 - 0.4\,z^{-1} - 0.32\,z^{-2}\big) = X(z)\big(1 + 0.1z^{-1}\big)$$

As $x(n) = \delta(n)$ therefore (from Table 9.1), $X(z) = 1$. the above equation can now be written as

$$Y(z) = \frac{\big(1 + 0.1z^{-1}\big)}{\big(1 - 0.4\,z^{-1} - 0.32\,z^{-2}\big)}$$

Multiplying both the numerator and the denominator with z^2, we get

$$Y(z) = \frac{z(z+0.1)}{\big(z^2 - 0.4z - 0.32\big)}$$

The denominator is factorized as

$$Y(z) = \frac{z(z+0.1)}{(z+0.4)(z-0.8)}$$

$$\Rightarrow \frac{Y(z)}{z} = \frac{(z+0.1)}{(z+0.4)(z-0.8)}$$

The right-hand side of the above equation is a proper rational polynomial, with the denominator polynomial having distinct poles. Therefore, it can be written into partial fractions as

$$\frac{Y(z)}{z} = \frac{A}{(z+0.4)} + \frac{B}{(z-0.8)}$$

To find out the unknown constants A and B, we use

$$A = \left[(z+0.4) \times \frac{X(z)}{z}\right]_{z=-0.4}$$

$$= \left[(z+0.4) \times \frac{(z+0.1)}{(z+0.4)(z-0.8)}\right]_{z=-0.4}$$

$$= \left[\frac{(z+0.1)}{(z-0.8)}\right]_{z=-0.4} = \frac{(-0.4+0.1)}{(-0.4-0.8)} = 0.25$$

$$B = \left[(z-0.8) \times \frac{X(z)}{z}\right]_{z=0.8}$$

$$= \left[(z-0.8) \times \frac{(z+0.1)}{(z+0.5)(z-0.8)}\right]_{z=0.8}$$

$$= \left[\frac{(z+0.1)}{(z+0.4)}\right]_{z=0.8} = \frac{(0.4+0.1)}{(0.4+0.4)} = 0.625$$

Equation (10.3) becomes

$$\frac{Y(z)}{z} = \frac{0.25}{(z+0.4)} + \frac{0.625}{(z-0.8)}$$

$$Y(z) = \frac{0.25\,z}{(z+0.4)} + \frac{0.625\,z}{(z-0.8)}$$

Taking an inverse z-transform of both sides

$$y(n) = Z^{-1}(Y(z))$$

$$= Z^{-1}\left(\frac{0.25\,z}{(z+0.4)}\right) + Z^{-1}\left(\frac{0.625\,z}{(z-0.8)}\right)$$

$$= (0.25)Z^{-1}\left(\frac{z}{(z-(-0.4))}\right) + (0.625)Z^{-1}\left(\frac{z}{(z-0.8)}\right)$$

$$= (0.25)(-0.4)^{n}\,u(n) + (0.625)(0.8)^{n}\,u(n)$$

Thus, the output signal is

$$y(n) = (0.25)(-0.4)^{n}\,u(n) + (0.625)(0.8)^{n}\,u(n)$$

10.3   POLE-ZERO DIAGRAMS FOR A FUNCTION IN THE Z-DOMAIN

The z-plane command computes and displays the pole-zero diagram of the z-function, as shown in Figure 10.8.

The command is

z plane(b,a)

To display the pole value, use root(a).
To display the zero value, use root(b).

$$X(z) = \frac{0.8\,z^{-1} + z^{-1}}{1 - 2\,z^{-1} + 3\,z^{-1}}$$

```
clc;
clear all
b=[0 0.8 1];
a= [1 -2 3];
roots(a)
roots(b)
zplane(b,a);
```

ans = 1.0000 + 1.4142i
1.0000 - 1.4142i
ans = -1.2500
Figure 10.8 shows the zero-pole diagram of the Z-function.

10.4   FREQUENCY RESPONSE USING Z-TRANSFORM

The Freqz function computes and displays the frequency response of the given z-transform of the function

freqz(b, a, Fs)

b = coeff. of numerator.
a = coeff. of denominator.
Fs = sampling frequency.

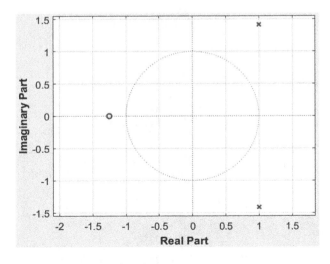

FIGURE 10.8   Zero-pole diagram of the z-function.

MATLAB Code:

```
clc;
clear all;
b=[2 10 4 5 3]
a= [5 3 2 1 1]
freqz(b,a);
```

Figure 10.9 shows the magnitude and phase response using *z*-transform.

Example 10.6

Plot the magnitude and phase of the frequency response of the given digital filter
Using freqz function:

$$Y(n) = 1.5\,x(n) + 0.75\,y(n-1) - 0.9\,y(n-2)$$

MATLAB Code:

```
clc;
clear all;
b = [1.5];
a= [1, -0.75, 0.9];
w = [0:1:100]*pi/100;
H=freqz(b,a,w);
magH = abs(H);
phaH = angle(H)*180/pi;
subplot(2,1,1);
plot(w/pi,magH);
title('Magnitude Response');
xlabel('frequency (rad/sec)');
ylabel('|H(f)|');
subplot(2,1,2);
plot(w/pi,phaH);
title('Phase Response');
xlabel('frequency (rad/sec)');
ylabel('Degrees');
```

Figure 10.10 shows magnitude and phase response using *z*-transform for Example 10.6.

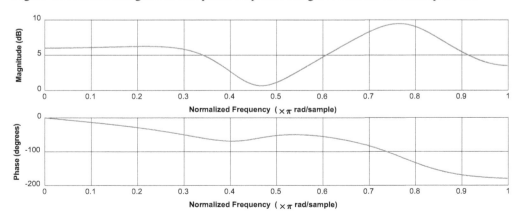

FIGURE 10.9 Magnitude and phase response **using *z*-transform.**

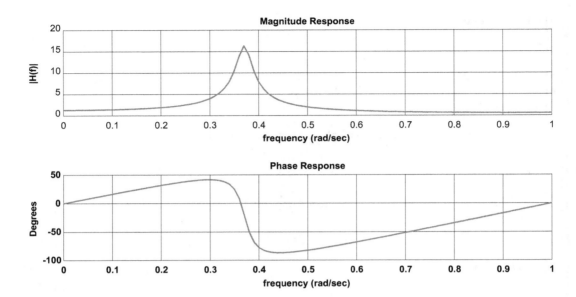

FIGURE 10.10   Magnitude and phase response **using z-transform for Example 10.6.**

PROBLEMS

10.1   Obtain the output for the following input $x(n)$ and impulse response $h(n)$ using z-transform
       $x(n) = 10\ u(n)$
       $h(n) = 5\ u(n)$

10.2   Find the impulse response and the transfer function of the following system as shown in Figure 10.11.

10.3   Find the impulse response and the transfer function of the following system as shown in Figure 10.12.

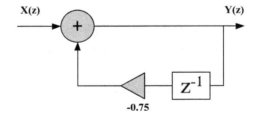

FIGURE 10.11   System for Problem 10.2.

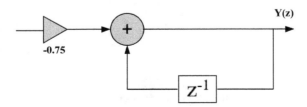

FIGURE 10.12   System for Problem 10.3.

10.4 A signal $x(n)$ begins at $n = 0$ and has seven finite sample values [1 2 3 2 1 –1 1] if it forms the input to a LTI processor whose impulse response $h(n)$ begins at $n = 0$ and has four finite sample values [1 1 1 1] convolute $x(n)$ with $h(n)$ to find output $y(n)$ using z-transform.

10.5 Given two sequences

$$x_1(n) = 5\delta(n) - 2\delta(n-2)$$

$$x_2(n) = 3\delta(n-3)$$

a. Determine the z-transform of convolution of the two sequences using the convolution property of the z-transform,

$$X(z) = X_1(z)X_2(z)$$

b. Determine convolution by the inverse z-transform from the result in part (a),

$$x(n) = Z^{-1}\left(X_1(z)X_2(z)\right)$$

10.6 Find the inverse z-transform for each of the following functions,

a. $X(z) = 4 - \dfrac{10z}{z-1} - \dfrac{z}{z+0.5}$

b. $X(z) = \dfrac{-5z}{(z-1)} + \dfrac{10z}{(z-1)^2} + \dfrac{2z}{(z-0.8)^2}$

c. $X(z) = \dfrac{z}{z^2 + 1.2z + 1}$

d. $X(z) = \dfrac{4z^{-4}}{z-1} + \dfrac{z^{-1}}{(z-1)^2} + z^{-8} + \dfrac{z^{-5}}{z-0.5}$

10.7 Using the partial fraction expansion method, find the inverse z-transform for each of the following functions,

a. $X(z) = \dfrac{1}{z^2 - 0.3z - 0.04}$

b. $X(z) = \dfrac{2}{(z-0.2)(z+0.4)}$

c. $X(z) = \dfrac{z}{(z+0.2)(z^2 - z + 0.5)}$

d. $X(z) = \dfrac{z(z+0.5)}{(z-0.1)^2(z-0.6)}$

10.8 A system is described by the difference equation

$$y(n) + 0.6y(n-1) = 4(0.8)^n u(n)$$

Determine the solution when the initial condition is $y(-1) = 2$.

10.9 A system is described by the difference equation

$$y(n) - 1.5y(n-1) + 0.06y(n-2) = 2(0.4)^{n-1}u(n-1)$$

Determine the solution when the initial condition is $y(-1) = 1$ and $y(-2) = 1$.

10.10 Given the following difference equation with the input–output relationship of a specific initially relaxed system (all initial conditions are zero),

$$y(n) - 0.7y(n-1) + 0.1y(n-2) = x(n) + 2x(n-1)$$

a. Find the impulse response sequence $y(n)$ due to the impulse sequence $\delta(n)$.
b. Find the output response of the system when the unit step function $u(n)$ is applied.

10.11 Given the following difference equation with the input–output relationship of a specific initially relaxed DSP system (all initial conditions are zero),

$$y(n) - 1.4y(n-1) + 0.29y(n-2) = x(n) + 1.5x(n-1)$$

a. Find the impulse response sequence $y(n)$ due to the impulse sequence $\delta(n)$.
b. Find the output response of the system when the unit step function $u(n)$ is applied.

10.12 Given the following difference equation,

$$y(n) = 0.5x(n) + 0.5x(n-1) + 0.5y(n-1)$$

a. Find the transfer function $H(z)$.
b. Determine the impulse response $y(n)$ if the input is $x(n) = 4\delta(n)$.
c. Determine the step response $y(n)$ if the input is $x(n) = 10u(n)$.

10.13 Given the following difference equation,

$$y(n) = x(n) - 0.5y(n-1) + 0.5y(n-2)$$

a. Find the transfer function $H(z)$.
b. Determine the impulse response $y(n)$ if the input is $x(n) = \delta(n)$.
c. Determine the step response $y(n)$ if the input is $x(n) = u(n)$.

10.14 Convert each of the following transfer functions into its difference equation

(a) $H(z) = \dfrac{z^2 - 0.25z}{z^2 + 1.1z + 0.18}$

(b) $H(z) = \dfrac{z^2 - 0.1z + 0.3}{z^3}$

10.15 Given the following digital system with a sampling rate of 10,000 Hz,

$$y(n) = 0.5x(n) + 0.5x(n-2)$$

a. Determine the frequency response of the system.
b. Calculate and plot the magnitude and phase-frequency responses.
c. Determine the filter type, based on the magnitude frequency response.

10.16 Given the following digital system with a sampling rate of 10,000 Hz,

$$y(n) = x(n) - 0.5y(n-2)$$

a. Determine the frequency response of the system.
b. Calculate and plot the magnitude and phase-frequency responses.
c. Determine the filter type, based on the magnitude frequency response.

10.17 Given the following difference equation for a digital system,

$$y(n) = x(n) - 2\cos(a)x(n-1) + x(n-2) + 2\gamma\cos(a) - \gamma^2$$

where $\gamma = 0.75$ and $a = 30°$,
a. Find the transfer function $H(z)$.
b. Plot the poles and zeros on the z-plane with the unit circle.
c. Determine the stability of the system from the pole-zero plot.

d. Calculate the amplitude (magnitude) frequency response of $H(z)$.

e. Calculate the phase-frequency response of $H(z)$.

10.18 Given the first-order IIR system

$$H(z) = \frac{1 - 2z^{-1}}{1 - 0.5z^{-1}}$$

Realize $H(z)$ and develop the difference equations using the following forms:

1. Direct-form I
2. Direct-form II

10.19 Given the filter

$$H(z) = \frac{1 - 0.9z^{-1} - 0.1z^{-2}}{1 - 0.3z^{-1} - 0.04z^{-2}}$$

Realize $H(z)$ and develop the difference equations using the following forms:

a. Direct-form I
b. Direct-form II
c. Cascade (series) form via the first-order sections
d. Parallel form via the first-order sections

10.20 Given the filter

$$H(z) = \frac{1 + 2z^{-1} + z^{-2}}{1 - 0.5z^{-1} + 0.25z^{-2}}$$

Use MATLAB to plot:

a. Its magnitude frequency response.
b. Its phase-frequency response.

10.21 Find the FFT for the input and output of the digital filter as shown in Figure 10.13 for the input signal $X(n) = [1\ 2\ 3\ 4]$.

10.22 Express the following z-transform in a factored form, plot its poles and zeros, and then determine its ROCs.

$$G(z) = \frac{4z^4 + 2z^3 + 4z^2 + 6z + 3}{3z^4 + 3z^3 - 1.5z^2 + z - 1.2}$$

10.23 Determine the partial fraction expansion of the z-transform $G(z)$ given by

$$G(z) = \frac{3z^3}{2z^3 + 3z^2 - 4z - 1}$$

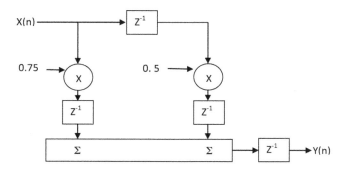

FIGURE 10.13 Filter for Problem 10.21.

11 Pole-Zero Stability

In DSP applications, the difference equation and transfer function are very important to study the characteristics of the system. The stability and frequency response can be examined based on the developed transfer function. This chapter will illustrate the concept of pole-zero stability, Stability determination based z-transform, determination pole and zeros from Difference Equation and Transfer Function and the Stability rules.

11.1 CONCEPT POLES AND ZEROS

A general causal digital filter has the difference equation:

$$y(n) = \sum_{i=0}^{N} a_i x(n-i) - \sum_{k=0}^{M} b_k x(n-k)$$

(11.1)

which is of the order $\max\{N,M\}$, and is recursive if any of the b_j coefficients are nonzero. A second-order recursive digital filter therefore has the difference equation:

$$y[n] = a_0\, x[n] + a_1\, x[n-1] + a_2\, x[n-2] - b_1\, y[n-1] - b_2\, y[n-2]$$

A digital filter with a recursive linear difference equation can have an infinite-impulse-response. Remember that the frequency response of a digital filter with impulse response $\{h[n]\}$ is:

$$H\left(e^{j\Omega}\right) = \sum_{n=-\infty}^{\infty} h(n)e^{-j\Omega n}$$

(11.2)

11.1.1 STABILITY DETERMINATION BASED Z-TRANSFORM

An LTI system that can be described using z-transform as a ratio is

$$H(z) = \frac{N(z)}{D(z)} = \frac{k\left(z-z_1\right)\left(z-z_2\right)\left(z-z_3\right)\cdots}{\left(z-p_1\right)\left(z-p_2\right)\left(z-p_3\right)\cdots}$$

(11.3)

Where k is the system gain and the constants $z1, z2, z3, \ldots$, are called zeros pf $X(z)$, because they are values of (z) for which $H(z)$ is zero. Conversely $p1, p2, p3, \ldots$, are called poles of $H(z)$. The poles and zeros are either a real or complex conjugate number.

The digital system is stable, if and only all poles of the system lie inside the unit circle in the z-plane.

11.1.2 THE Z-TRANSFORM

Consider the response of a causal stable LTI digital filter to the special sequence $\{z^n\}$ where z is complex. If $\{h[n]\}$ is the impulse response, by discrete-time convolution, the output is a sequence $\{y[n]\}$ where

$$y(n) = \sum_{k=-\infty}^{\infty} h(n)z^{n-k} = z^n \sum_{k=-\infty}^{\infty} h(n)z^{-k}$$

(11.4)

$$= z^n H(z)$$

The expression obtained for $H(z)$ is the "z-transform" of the impulse response. $H(z)$ is a complex number when evaluated for a given complex value of z.

It indicates that for a stable causal system, $H(z)$ must be finite when evaluated for a complex number z with modulus greater than or equal to one.

$$\text{Since } H(z) = \sum_{n=-\infty}^{\infty} h[n]z^{-n} \text{ and the frequency-response} : H(e^{i\Omega}) = \sum_{n=-\infty}^{\infty} h[n]z^{-j\Omega n}$$

it is clear that replacing z by $e^{j\Omega}$ in $H(z)$ gives $H(e^{j\Omega})$.

11.1.3 THE "z-PLANE"

Conveniently, it is represented as complex numbers on an "Argand diagram" as illustrated in Figure 11.1. The main reason for doing this is that the modulus of the difference between two complex numbers $a + jb$ and $c + jd$ say, i.e., $| (a + jb) - (c + jd) |$ is represented graphically as the length of the line between the two complex numbers as plotted on the Argand diagram.

If one of these complex numbers, $c + jd$ say is zero, i.e., $0 + j0$, then the modulus of the other number $|a + jb|$ is the distance of $a + jb$ from the origin $0 + j0$ on the Argand diagram.

Of course, any complex number like $a + jb$ can be converted into polar form $Re^{j\theta}$ where $R = |a + jb|$ and $\theta = \tan^{-1}(b/a)$. Plotting a complex number expressed as $Re^{j\theta}$ on an Argand diagram is also illustrated above. We draw an arrow of length R starting from the origin and set at an angle θ from the "real part" axis (measured anti-clockwise). $Re^{j\theta}$ is then at the tip of the arrow. In the illustration above, θ is about $\pi/4$ or 45 degrees. If $R = 1$, $Re^{j\theta} = e^{j\theta}$ and on the Argand diagram would be a point at a distance one from the origin. Plotting $e^{j\theta}$ for values θ in the range of 0 to 2π 360° produces points all of which lie on a "unit circle," i.e., a circle of radius 1, with the center as the origin.

Where the complex numbers plotted on an z-diagram are values of z for which we are interested in $H(z)$, the diagram is referred to as "the z-plane." Points with $z = e^{j\Omega}$ lie on a unit circle, as shown in Figure 11.2. Remember that $|e^{j\Omega}| = |\cos(\Omega) + j\sin(\Omega)| = \sqrt{[\cos^2(\Omega) + \sin^2(\Omega)]} = 1$. Therefore, evaluating the frequency response $H(e^{j\Omega})$ for Ω in the range 0 to π is equivalent to evaluating $H(z)$ for $z = e^{j\Omega}$ which goes around the upper part of the unit circle as Ω goes from 0 to π.

11.2 DIFFERENCE EQUATION AND TRANSFER FUNCTION

The general difference equation is given by

$$y(n) = b_0 x(n) + b_0 x(n-1) + \cdots + b_M x(n-M) - a_1 y(n-1) + \cdots - a_N y(n-N) \quad (11.5)$$

The supposition that all initial conditions of this system are zero, and the $X(z)$, and $Y(z)$ represent the z-transforms of the sequences $x(n)$, and $y(n)$, respectively, taking the z-transform of Equation (11.5) yields

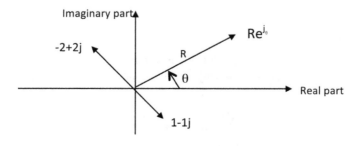

FIGURE 11.1   Complex number phasor diagram.

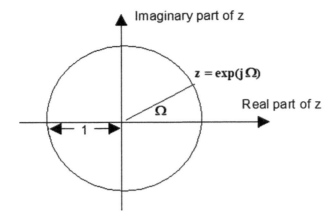

FIGURE 11.2   z-plane diagram.

$$Y(z) = b_0 X(z) + b_0 X(z)z^{-1} + \cdots + b_M X(z)z^{-M} - a_1 Y(z)z^{-1} + \cdots - a_N Y(z)z^{-N} \tag{11.6}$$

The z-transfer function $H(z)$ defined as the ratio of z-transform of the output $Y(z)$ to the z-transform of the input $X(z)$. To get the transfer function rearrange Equation (11.6) as

$$H(z) = \frac{Y(z)}{X(z)} = \frac{b_0 + b_1 z^{-1} + \cdots + b_M z^{-M}}{1 + a_1 z^{-1} + \cdots + a_N z^{-N}} = \frac{A(z)}{B(z)} \tag{11.7}$$

where $H(z)$ defined as the transfer function with its numerator and denominator polynomials is defined as

$$A(z) = b_0 + b_1 z^{-1} + \cdots + b_M z^{-M} \tag{11.8}$$

$$B(z) = 1 + a_1 z^{-1} + \cdots + a_N z^{-N} \tag{11.9}$$

The z-transfer function represents the digital filter in the z-domain, as shown in Figure 11.3.

Example 11.1

A DSP system is described by the following difference equation

$$y(n) = x(n) + x(n-1) - 1.2x(n-2) - 2y(n-1) - 0.8y(n-2)$$

Find the transfer function $H(z)$, the numerator polynomial $A(z)$, and the denominator polynomial equation $B(z)$.

SOLUTION

Taking the z-transform on both sides of the previous difference equation we achieve
Moving the last two terms to the left side of the difference equation and factoring $Y(z)$, i.e.
$-2y(n-1) - 0.8y(n-2)$

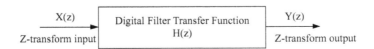

FIGURE 11.3   z-transfer function and represent as the digital filter.

On the left side $X(z)$ on the right side, we obtain

$$y(n) = x(n) + x(n-1) - 1.2x(n-2) - 2y(n-1) - 0.8y(n-2)$$

$$Y(z) = X(z) + X(z)Z^{-1} - 1.2X(z)Z^{-2} - 2Y(z)Z^{-1} - 0.8Y(z)Z^{-2}$$

$$Y(z) + 2Y(z)Z^{-1} + 0.8Y(z)Z^{-2} = X(z) + X(z)Z^{-1} - 1.2X(z)Z^{-2}$$

$$\left[1 + 2Z^{-1} + 0.8Z^{-2}\right]Y(z) = \left[1 + Z^{-1} - 1.2Z^{-2}\right]X(z)$$

$$\frac{Y(z)}{X(z)} = \frac{1 + Z^{-1} - 1.2Z^{-2}}{1 + 2Z^{-1} + 0.8Z^{-2}}$$

$$\frac{Y(z)}{X(z)} = \frac{Z^2 + Z - 1.2}{Z^2 + 2Z + 0.8}$$

The numerator polynomial $A(z) = Z^2 + Z - 1.2$
The denominator polynomial $B(z) = Z^2 + 2Z + 0.8$

Example 11.2

A DSP system is described by the following difference equation

$$y(n) = x(n) - 0.35x(n-1) + 1.2x(n-2)$$

Find the transfer function $H(z)$, the numerator polynomial $A(z)$, and the denominator polynomial equation $B(z)$.

SOLUTION

Taking the z-transform on both sides of the previous difference equation we achieve

$$y(n) = x(n) - 0.35x(n-1) + 1.2x(n-2)$$

$$Y(z) = X(z) - 0.35X(z)Z^{-1} + 1.2X(z)Z^{-2}$$

$$Y(z) = \left[1 - 0.35Z^{-1} + 1.2Z^{-2}\right]X(z)$$

$$\frac{Y(z)}{X(z)} = \left[1 - 0.35Z^{-1} + 1.2Z^{-2}\right]$$

The numerator polynomial is $A(z) = 1 - 0.35Z^{-1} + 1.2Z^{-2}$
The denominator polynomial is $B(z) = 1$

11.3   BIBO STABILITY

A system is said to be bounded-input/bounded-output stable (BIBO stable or just stable) if the output signal is bounded for all input signals that are bounded.

Consider a discrete-time system with input x and output y. The input is said to be bounded if there is a real number $M < \infty$ such that $|x(k)| \leq M$ for all k.

An output is bounded if there is a real number $N < \infty$ such that $|y(k)| \leq N$ for n.

The system is stable if, for any input bounded by M, there is some bound N on the output.

Theorem:
A discrete-time LTI system is stable if and only if its impulse response is summable.

Proof:

Consider a discrete-time LTI system with impulse response h. The output y corresponding to the input x is given by the convolution sum,

$$\forall n \in \text{Integers}, y(n) = \sum_{m=-\infty}^{\infty} h(m)x(n-m) \tag{11.10}$$

Suppose that the input bounded with bound M. Then, applying the triangle inequality, we see that

$$|y(n)| \leq \sum_{m=-\infty}^{\infty} |h(m)||x(n-m)| \leq M \sum_{m=-\infty}^{\infty} |h(m)| \tag{11.11}$$

Thus, if the impulse response is summable, then the output is bounded with bound

$$N = M \sum_{m=-\infty}^{\infty} |h(m)| \tag{11.12}$$

Proof:

To show that the system is not stable, we need to find one bounded input for which the output either does not exist or is not bounded. Such information is given by

$$\forall n \in \text{Integers},$$

$$x(n) = \frac{h(-n)}{|h(-n)|}, h(n) \neq 0$$
$$= 0, h(n) = 0 \tag{11.13}$$

The input is bounded, with bound $M = 1$. Plugging this input to the convolution sum (1) and evaluating at $n = 0$, we get

$$y(0) = \sum_{m=-\infty}^{\infty} h(m)x(-m) = \sum_{m=-\infty}^{\infty} \frac{(h(m))^2}{|h(m)|} = \sum_{m=-\infty}^{\infty} |h(m)| \tag{11.14}$$

But since the impulse response is not summable, $y(0)$ does not exist or is not finite, so the system is not stable.

11.4   THE Z-PLANE POLE-ZERO PLOT AND STABILITY

A handy tool for analyzing digital systems is the z-plane pole-zero plot. This graphical technique allows us to investigate the characteristics of the digital system shown in Figure 11.3, including system stability. In general, a digital transfer function can be written in the pole-zero form, and we can plot the poles and zeros on the z-plane. The z-plane is depicted in Figure 11.4 and has the following features:

1. The horizontal axis is the real part of the variable z, and the vertical axis represents the imaginary part of the variable z.
2. The z-plane is divided into two parts by a unit circle.
3. Each pole is marked on the z-plane using the cross symbol ×, while each zero is plotted using the small circle symbol.

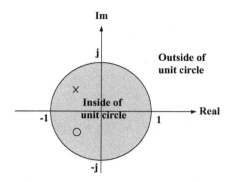

FIGURE 11.4   *z*-plane pole-zero plot.

11.5   STABILITY RULES

Similar to the analog system, the digital system requires that all poles plotted on the *z*-plane must be inside the unit circle. We summarize the rules for determining the stability of a DSP system as follows:

1. If the outermost pole(s) of the *z*-transfer function $H(z)$ describing the DSP system is(are) inside the unit circle in the *z*-plane pole-zero plot, then the system is stable.
2. If the outermost pole(s) of the *z*-transfer function $H(z)$ is(are) outside the unit circle in the *z*-plane pole-zero plot, the system is unstable.
3. If the outermost pole(s) is(are) first-order pole(s) of the *z*-transfer function $H(z)$ and on the unit circle in the *z*-plane pole-zero plot, then the system is marginally stable.
4. If the outermost pole(s) is(are) multiple-order pole(s) of the *z*-transfer function $H(z)$ and on the unit circle in the *z*-plane pole-zero plot, then the system is unstable.
5. The zeros do not affect system stability.

Notice that the following facts apply to a stable system (bounded-in/bounded-out [BIBO] stability discussed in Chapter 2):

1. If the input to the system is bounded, then the output of the system will also be bounded, or the impulse response of the system will go to zero in a finite number of steps.
2. An unstable system is one in which the output of the system will grow without bound due to any bounded input, initial condition, or noise, or its impulse response will grow without being bound.
3. The impulse response of a marginally stable system stays at a constant level or oscillates between two finite values.

Example 11.3

The example illustrating the rules of stability
When the input is impulse sequence $x(n) = \delta(n)$ as in Figure 11.5(a) and the difference output equation is given as $y(n) = x(n) + 0.75y(n-1)$, the transfer function is given as

$$H(z) = \frac{z}{z - 0.75} = Y(z)$$

and the output response can be written as $y(n) = (0.75)^n u(n)$ as in Figure 11.5(b).
The *z*-plane is given in Figure 11.5(c) and shows the stability of the system.

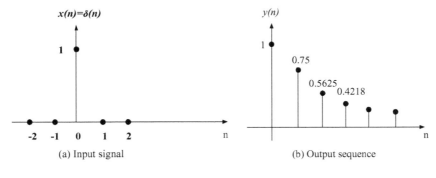

(a) Input signal   (b) Output sequence

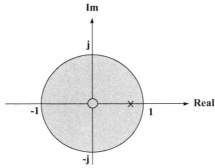

(c) Z-plane representation

FIGURE 11.5   Response of a stable system. (a) Input signal. (b) Output sequence. (c) Z-plane representation.

When the input is an impulse sequence $x(n) = \delta(n)$ as in Figure 11.6(a) and the difference output equation is given as $y(n) = x(n) + 1.25y(n-1)$, the transfer function is given as

$$H(z) = \frac{z}{z - 1.25} = Y(z)$$

And the output response can be written as $y(n) = (1.25)^n u(n)$ as in Figure 11.6b

The z-plane is given in Figure 11.6c and shows the stability of the system.

When the input is an impulse sequence $x(n) = \delta(n)$ as in Figure 11.7 and the difference output equation is given as $y(n) = x(n) + y(n-1)$, the transfer function is given as

$$H(z) = \frac{z}{z - 1} = Y(z)$$

And the output response can be written as $y(n) = u(n)$ as in Figure 11.8.

The z-plane is given in Figure 11.9 and shows the stability of the system.

Example 11.4

Given the following transfer function

$$H(z) = \frac{z^{-1} - 0.6z^{-2}}{1 + 1.2z^{-1} + 0.55z^{-2}}$$

Convert it into its pole-zero form.

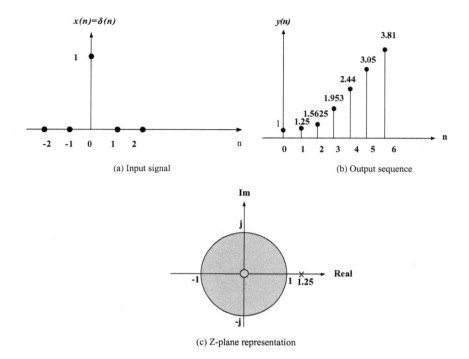

FIGURE 11.6 Response of the unstable system. (a) Input signal. (b) Output sequence. (c) Z-plane representation.

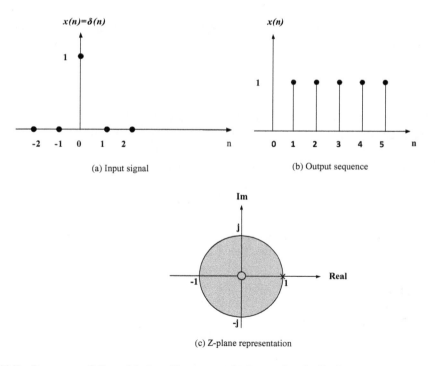

FIGURE 11.7 Response of the critical stable system. (a) Input signal. (b) Output sequence. (c) Z-plane representation.

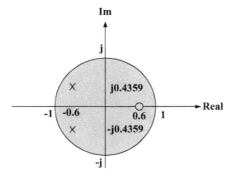

FIGURE 11.8 z-plane representation of Example 11.4.

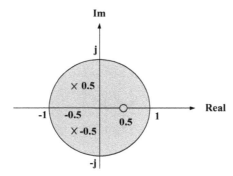

FIGURE 11.9 z-plane representation of Example 11.6a.

SOLUTION

We first multiply the numerator and denominator by z^2 to obtain the transfer function whose numerator and the denominator polynomials have the positive powers of z, as follows

$$H(z) = \frac{z^2\left(z^{-1} - 0.6z^{-2}\right)}{z^2\left(1 + 1.2z^{-1} + 0.55z^{-2}\right)}$$

$$H(z) = \frac{z - 0.6}{z^2 + 1.2z + 0.55}$$

The zero of $H(z)$: $z - 0.6 = 0 \rightarrow z = 0.6$
The poles of $H(z)$: $z^2 + 1.2z + 0.55 = 0 \rightarrow z = -0.6 \pm j0.4359$

From Figure 11.10, the poles are seen inside the unit circle, so the system is stable

Example 11.5

Given the following transfer function

$$H(z) = \frac{z\left(1 - z^{-2}\right)}{\left(1 + 1.3z^{-1} + 0.36z^{-2}\right)}$$

Convert it into its pole-zero form.

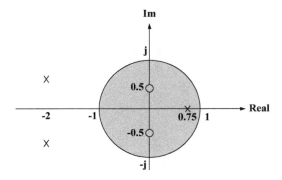

FIGURE 11.10 z-plane representation of Example 11.4b.

SOLUTION

We first multiply the numerator and denominator by z^2 to obtain the transfer function whose numerator and the denominator polynomials have the positive powers of z, as follows

$$H(z) = \frac{\left(1-z^{-2}\right)z^2}{\left(1+1.3z^{-1}+0.36z^{-2}\right)z^2} = \frac{z^2-1}{z^2+1.3z+0.36}$$

Putting the numerator polynomial equal to zero and then finding the roots, gives us the zeros of the transfer function,

$$z^2 - 1 = 0$$

$$(z-1)(z+1) = 0$$

Therefore, we get $z_1 = 1$ and $z_2 = -1$ as the roots.

Now, setting the denominator polynomial equal to zero and find the roots, gives us the poles of the transfer function,

$$z^2 + 1.3z + 0.36 = 0$$

$$z = \frac{-1.3 \pm \sqrt{(1.3)^2 - 4(1)(0.36)}}{2(1)} = \frac{-1.3 \pm \sqrt{1.69 - 1.44}}{2}$$

$$= \frac{-1.3 \pm \sqrt{0.25}}{2} = \frac{-1.3 \pm 0.5}{2} = -0.4, -0.9$$

Therefore, the poles are $p_1 = -0.4$ and $p_{21} = -0.9$. The transfer function can now be written in the pole-zero form as

$$H(z) = \frac{(z-1)(z+1)}{(z+0.4)(z+0.9)}$$

Example 11.6

The following transfer functions describe digital systems

a) $H(z) = \dfrac{(z-0.5)}{(z-0.5)(z^2+z+0.5)}$

b) $H(z) = \dfrac{\left(z^2 + 0.25\right)}{\left(z - 0.5\right)\left(z^2 + 4z + 4.5\right)}$

c) $H(z) = \dfrac{\left(z + 0.25\right)}{\left(z - 0.25\right)\left(z^2 + 1.5z + 1\right)}$

d) $H(z) = \dfrac{\left(z^2 + z + 0.25\right)}{\left(z - 1\right)^2 \left(z + 1\right)\left(z - 0.6\right)}$

For each, sketch the z-plane pole-zero plot and determine the stability status for the digital system.

SOLUTION

a) Put the numerator polynomial equal to zero and then find the roots,

$$z - 0.5 = 0$$

Therefore, we get $z_1 = 0.5$ as the root.
Now, set the denominator polynomial equal to zero and find the roots, which gives us the poles of the transfer function,

$$\left(z - 0.5\right)\left(z^2 + z + 0.5\right) = 0$$

This leads to

$$z - 0.5 = 0$$

and

$$z = \frac{-1 \pm \sqrt{(1)^2 - 4(1)(0.5)}}{2(1)} = \frac{-1 \pm \sqrt{1 - 2}}{2}$$

$$= \frac{-1 \pm \sqrt{-1}}{2} = \frac{-1.0 \pm j}{2} = -0.5 \pm j0.5$$

Therefore, the poles are $p_1 = 0.5$, $p_2 = -0.5 + j0.5$ and $p_3 = -0.5 - j0.5$. The magnitudes of these poles are

$$\left|p_1\right| = 0.5$$

$$\left|p_2\right| = \left|-0.5 + j0.5\right| = \sqrt{(-0.5)^2 + (0.5)^2} = 0.707$$

$$\left|p_3\right| = \left|-0.5 - j0.5\right| = \sqrt{(-0.5)^2 + (-0.5)^2} = 0.707$$

It can be noticed that the magnitudes of all the poles are less than 1, so they are inside the unit circle in the z-plane pole-zero plot. Therefore, the system is stable. It is shown in the Figure 11.9.

b) Put the numerator polynomial equal to zero and then find the zero roots of the transfer function,

$$z^2 + 0.25 = 0$$

$$z^2 = -0.25$$

$$z = \sqrt{-0.25} = \pm j0.5$$

Therefore, we get $z_1 = +j0.5$ and $z_2 = -j0.5$ as the roots.

Now, set the denominator polynomial equal to zero and find the roots, which gives us the poles of the transfer function,

$$(z - 0.75)(z^2 + 4z + 4.5) = 0$$

It leads to

$$z - 0.75 = 0 \rightarrow z = 0.75$$

and

$$z = \frac{-4 \pm \sqrt{(4)^2 - 4(1)(4.5)}}{2(1)} = -2 \pm j0.707$$

Therefore, the poles are $p_1 = 0.75$, $p_2 = -2 + j0.707$, and $p_3 = -2 - j0.707$. It can be noticed that the two poles are outside the unit circle on z-plane pole-zero. Therefore, the system is unstable. It is shown in Figure 11.10.

c) Put the numerator polynomial equal to zero and then find the roots,

$$z + 0.25 = 0$$

Therefore, we get $z_1 = -0.25$ as the root.
Now, setting the denominator polynomial equal to zero and finding the roots, gives us the poles of the transfer function,

$$(z - 0.25)(z^2 + 1.5z + 1) = 0$$

This leads to

$$z - 0.5 = 0 \text{ so } z = 0.25$$

and

$$z = \frac{-1.5 \pm \sqrt{(1.5)^2 - 4(1)(1)}}{2(1)} = -0.75 \pm j0.66$$

Therefore, the poles are $p_1 = 0.25$, $p_2 = -0.75 + j0.66$ and $p_3 = -0.75 - j0.66$. The magnitudes of these poles are

$$|p_1| = 0.25$$

$$|p_2| = |-0.75 + j0.66| = 1$$

$$|p_3| = |-0.75 - j0.66| = 1$$

It can be noticed that the magnitudes of the two poles are on unity circle 1, so the system is marginally stable. This is shown in Figure 11.9.

d) $(z^2 + z + 0.25) = 0$ gives the two zeros at $z = -0.5$

$$(z - 1)^2 (z + 1)(z - 0.6) = 0$$

gives the poles at $p = 1$ (two poles) and one pole at $p = -1$ and $p = 0.6$, respectively And the system is marginally stable.

Example 11.7

Check the stability of the system given by

$$H(z) = \frac{A(z - 1)^2}{(z - 0.4)(z - 0.5 + j0.5)(z - 0.5 - j0.5)}$$

where A is a constant number.

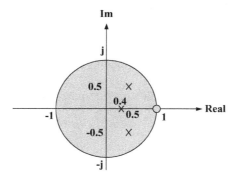

FIGURE 11.11 z-plane representation of Example 11.7.

SOLUTION

$$z - 0.4 = 0 \rightarrow z = 0.4$$

$$z - 0.5 + j0.5 = 0 \rightarrow z = 0.5 - j0.5$$

$$z - 0.5 - j0.5 = 0 \rightarrow z = 0.5 + j0.5$$

The poles are inside the unit circle, so the system is stable (Figure 11.11).

PROBLEMS

11.1 Convert each of the following transfer functions into its pole-zero form.

$$H(z) = \frac{z\left(1 - 0.16z^{-2}\right)}{1 + 0.7z^{-1} + 0.1z^{-2}}$$

11.2 Given that each of the following transfer functions describe digital systems, sketch the z-plane pole-zero plot, and determine the stability status for the digital system.

(a) $H(z) = \dfrac{z(z - 0.75)}{(z + 0.25)(z^2 + z + 0.8)}$

(b) $H(z) = \dfrac{z(z^2 + 0.25)}{(z - 0.5)(z^2 + 4z + 7)}$

(c) $H(z) = \dfrac{z(z + 0.15)}{(z + 0.2)(z^2 + 1.4141z + 1)}$

(d) $H(z) = \dfrac{z(z^2 + z + 0.25)}{(z - 1)(z + 1)^2(z - 0.36)}$

11.3 An LTI system is represented by the difference equation:

$$y(n) = 0.7\, y(n - 1) - 0.12\, y(n - 2) + x(n - 1) + x(n - 2)$$

1. Find the system transfer function $H(z)$.
2. Draw and obtain the poles and zeros; is the system is stable?
3. Find the output response if the input is the unit step sequence given by $x(n) = u(n)$.

11.4   A DSP system is described by the following difference equation

$$y(n) = x(n) - 0.9x(n-1) + 1.8x(n-2)$$

Find the transfer function $H(z)$, the numerator polynomial $A(z)$, and the denominator polynomial equation $B(z)$.

11.5   A DSP system is described by the following difference equation

$$y(n) = x(n) - 1.2x(n-1) + 0.7x(n-2)$$

Find the transfer function $H(z)$, the numerator polynomial $A(z)$, and the denominator polynomial equation $B(z)$.

11.6   The following transfer functions describe digital systems.

(a) $H(z) = \dfrac{(z - 1.5)}{(z + 1.5)(z^2 + z + 0.5)}$

(b) $H(z) = \dfrac{(z^2 + 1.25)}{(z - 0.5)(z^2 + 4z + 3)}$

(c) $H(z) = \dfrac{(z + 0.25)}{(z - 0.25)(z^2 + 1.5z + 3)}$

(d) $H(z) = \dfrac{(z^2 + z + 0.25)}{(z - 1)^2 (z + 0.9)(z - 0.2)}$

For each, sketch the z-plane pole-zero plot and determine the stability status for the digital system.

12 Sampling

In digital signal processing, sampling is the convert a continuous-time signal to a discrete-time signal, Such as conversion of a sound wave (a continuous signal) to a sequence of samples (a discrete-time signal). A sample is a value or set of values at a point in time and space. A sampler is a subsystem or operation that extracts samples from a continuous signal. A theoretical ideal sampler produces samples equivalent to the instantaneous value of the continuous signal at the desired points.

12.1 RELATING THE FT TO THE DTFT FOR DISCRETE-TIME SIGNALS

First, we must establish the relation between discrete-time-frequency Ω_0 and continuous-time-frequency ω_0.

Given,

$$x(t) = e^{j\omega t} \quad \text{and} \quad g(n) = e^{j\Omega n} \tag{12.1}$$

Let, $g(n) = x(nT)$, this implies that,

$$e^{j\Omega n} = e^{j\omega Tn} \tag{12.2}$$

and we may define $\Omega = \omega T$.

Now, to relate FT to DTFT consider:

$$X\left(e^{j\Omega}\right) = \sum_{n=-\infty}^{\infty} x(n) e^{-j\Omega n} \tag{12.3}$$

The aim is to seek an FT pair $x_s(t) \overset{FT}{\Leftrightarrow} X_s(j\omega)$ which corresponds to $x(n) \overset{DTFT}{\Leftrightarrow} X\left(e^{j\Omega}\right)$

where $x_s(t)$ is a continuous-time signal that *corresponds* to $x(n)$, and the FT $X_s(j\omega)$ corresponds to the DTFT $X\left(e^{j\Omega}\right)$.

Now,

$$X_s(j\omega) = X\left(e^{j\Omega}\right)\Big|_{\Omega=\omega T} \tag{12.4}$$

$$X_s(j\omega) = \sum_{n=-\infty}^{\infty} x(n) e^{-j\omega Tn} \tag{12.5}$$

Taking inverse FT for $X_s(j\omega)$ by using

$$\delta(t - nT) \overset{FT}{\Leftrightarrow} e^{j\omega Tn} \tag{12.6}$$

To obtain the continuous-time signal description of the discrete-time signal:

$$x_s(t) = \sum_{n=-\infty}^{\infty} x(n) \delta(t - nT) \tag{12.7}$$

and, hence,

$$x_s(t) = \sum_{n=-\infty}^{\infty} x(n)\, \delta(t - nT) \overset{FT}{\Leftrightarrow} X_s(j\omega) = \sum_{n=-\infty}^{\infty} x(n)\, e^{-j\omega Tn} \tag{12.8}$$

See the following (Figure 12.1).

12.2 SAMPLING

The sampling operation generates a discrete-time signal $x(n)$ from a continuous-time signal $x(t)$. Note that $x(t)$ is a band-limited signal.

12.3 BAND-LIMITED SIGNALS

A band-limited signal is a signal $g(t)$ with a spectrum which is zero above a specified frequency f_H Hz.

$$g(t) \Leftrightarrow G(\omega) = 0 \quad \text{for} \quad |\omega| > \omega_H = 2\pi f_H \tag{12.9}$$

12.4 SAMPLING OF CONTINUOUS-TIME SIGNALS

Let $x(n)$ be a discrete-time signal that is equal to the samples of $x(t)$ at integer multiples of a sampling interval T. That is $x(n) = x(nT)$. The effect of sampling is evaluated by relating the DTFT of $x(n)$ to the FT of $x(t)$. Figure 12.2 shows the concept of sampling.

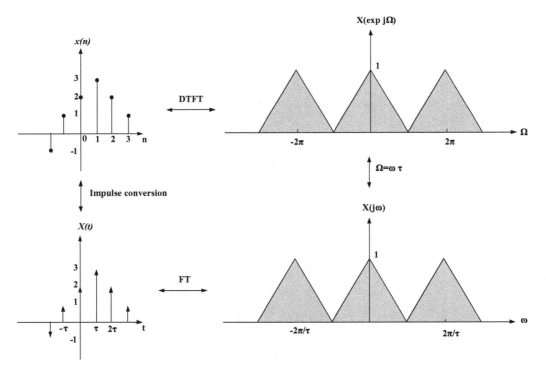

FIGURE 12.1 Relation between FT and DTFT.

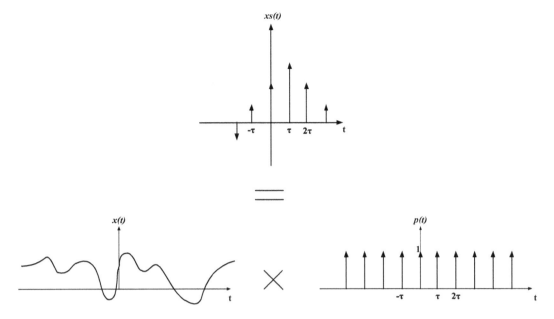

FIGURE 12.2 Concept of sampling.

Begin with the continuous-time representation for the discrete-time signal $x(n)$:

$$x_s(t) = \sum_{n=-\infty}^{\infty} x(n)\,\delta(t - nT) \tag{12.10}$$

$$x_s(t) = \sum_{n=-\infty}^{\infty} x(nT)\,\delta(t - nT) \tag{12.11}$$

Since, $x(t)\,\delta(t - nT) = x(nT)\,\delta(t - nT)$, we may rewrite $x_s(t)$ as:

$$x_s(t) = x(t)p(t) \tag{12.12}$$

where

$$p(t) = \sum_{n=-\infty}^{\infty} \delta(t - nT) \tag{12.13}$$

Then, multiplication in the time domain corresponds to convolution in the frequency domain:

$$X_s(j\omega) = \frac{1}{2\pi}\, X(j\omega) * P(j\omega) \tag{12.14}$$

$$X_s(j\omega) = \frac{1}{2\pi}\, X(j\omega) * \frac{2\pi}{T} \sum_{k=-\infty}^{\infty} \delta(\omega - k\omega_0) \tag{12.15}$$

Where, $\omega_s = \dfrac{2\pi}{T}$ is the sampling frequency.

Now, convolve $X(j\omega)$ with each impulse:

$$X_s(j\omega) = \frac{1}{T} \sum_{k=-\infty}^{\infty} X(\omega - k\omega_s) \qquad (12.16)$$

Notes on the above equations:

1. The FT of the sampled signal is given by an infinite sum of versions of the original signal's FT.
2. The shifted versions are offset by integer multiples of ω_s.
3. The shifted versions may overlap with each other if ω_s is not large enough compared to the frequency content of $X(j\omega)$.

Assume that $x(t)$ has frequency contents of $-Wn < \omega < Wp$. Under this condition, three cases are stated:

1. $\omega_s = 3W$. See Figure 12.3(b).
2. $\omega_s = 2W$. See Figure 12.3(c).
3. $\omega_s = \dfrac{3}{2}W$. See Figure 12.3(d).

We conclude that the shifted versions of $X(j\omega)$ overlap one another when $\omega_s < 2W$.
Aliasing phenomena (see Figure 12.3 (d))

- Aliasing is the overlapping in the shifted replicas of the original signal.
- It occurs when $\omega_s < 2W$.
- It distorts the spectrum of the original signal.
- Therefore, the original signal spectrum $X(j\omega)$ (the middle one in Figure 12.3 (d)) *cannot* be reconstructed.
- To prevent aliasing, we must have $\omega_s > 2W$, where W is the highest frequency component in the signal.

12.5 SAMPLING THEOREM

The sampling theorem states that a band-limited signal which has no frequency components higher than f_H Hz can be recovered completely from a set of samples taken at the rate of f_s $(\geq 2f_H)$ samples per second.

The above sampling theorem is also called the uniform sampling theorem for *base-band* or *low-pass signals*.

The minimum sampling rate, $2f_H$ samples per second, is called the Nyquist rate; its reciprocal $1/2f_H$, measured in seconds, is called the Nyquist interval.

In sampling, a strict distinction between base-band and pass-band signals made as follows:

For *base-band* signals, $B \geq f_L$ H_z.
For *pass-band* signals, $B < f_L$ H_z.

where B is the bandwidth of the signal $g(t)$, and f_L is the lowest frequency of the signal $g(t)$.

Sampling Theory:

The sample concept of sampling is given as switch as shown in Figure 12.4, and the sampled signal using switching sampling is given in Figure 12.5.

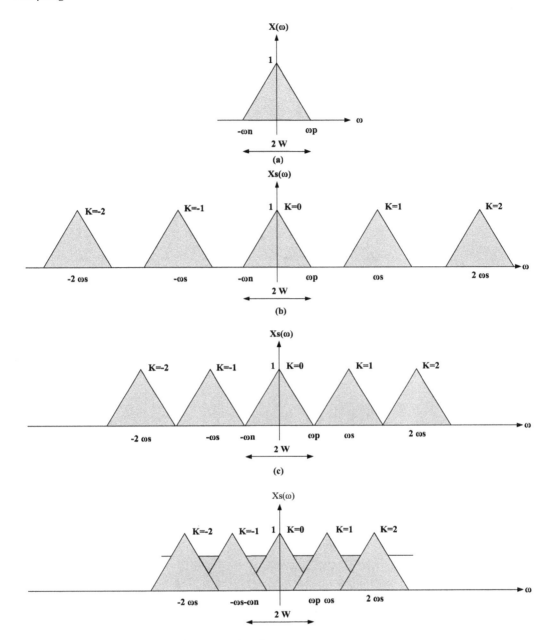

FIGURE 12.3 Sampling at a different sampling frequency ω_s.

fs = sampling frequency rate in sample/sec or Hz.

$$T_s = \frac{1}{f_s} \tag{12.18}$$

T_s sampling period time (sec).

Nyquist Condition

$$f_s \geq 2f_m \tag{12.19}$$

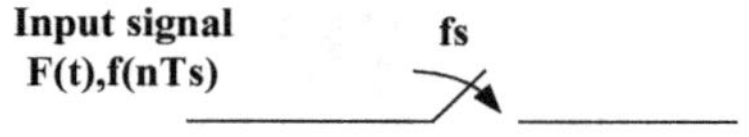

Input signal
F(t),f(nTs)   **fs**

FIGURE 12.4   Basic concept of sampling.

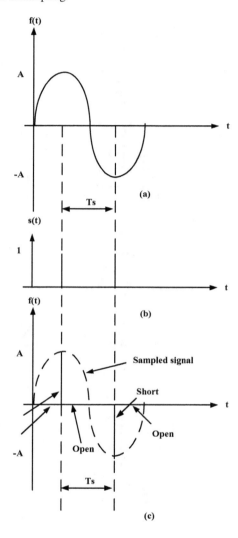

FIGURE 12.5   Sampled signal using switching sampling.

Nyquist rate frequency in sample/sec or Hz.

$$f_{s(min)} = 2f_m \qquad\qquad (12.20)$$

Example 12.1

Calculate the sampling rate frequency for the signal $f(t) = 2\sin(2\pi \times 10^3 t)$ (Figure 12.6).

SOLUTION

$$\omega_m = 2\pi \times 10^3 \frac{\text{rad}}{\text{sec}} = 2\pi f_m$$

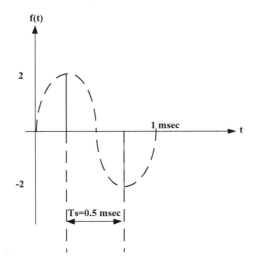

FIGURE 12.6 Sample signal using switching sampling of Example 12.1.

$$f_m = 1\text{kHz} \rightarrow T_m = \frac{1}{f_m} = 1\text{msec}$$

$$f_s = 2f_m \rightarrow f_s = 2 \times 1 = 2\text{msec}$$

$$T_s = \frac{1}{f_s} = 0.5\text{msec}$$

Example 12.2

Find the Nyquist rate and the Nyquist interval for each of the following signals:

(a) $f(t) = 2\cos(800\pi t)\cos(4200\pi t)$

(b) $f(t) = \dfrac{\sin(300\pi t)}{\pi t}$

(c) $f(t) = \left(\dfrac{\cos(250\pi t)}{\pi t}\right)^2$

SOLUTION

(a) $f(t) = 2\cos(800\pi t)\cos(4200\pi t)$

$= \cos(3400\pi t) + \cos(5000\pi t)$

Thus $f(t)$ is a band-limited signal with $f_H = 2{,}500$ Hz. Hence the Nyquist rate is 5,000 Hz and the Nyquist interval is $\dfrac{1}{5000} = 0.2\text{ms}$.

(b) $f(t) = \dfrac{\sin(300\pi t)}{\pi t} \Leftrightarrow \dfrac{1}{300}\,\text{rect}\left(\dfrac{\omega}{600\pi}\right)$

Thus $f(t)$ is a band-limited signal with $\omega_H = 300\pi$ (i.e., $f_H = 150$ Hz). Hence, the Nyquist rate is 300 Hz and the Nyquist interval is $\dfrac{1}{300} = 3.33\text{ms}$.

(c) From the frequency convolution theorem, we find that the signal $f(t)$ is also band-limited and its bandwidth is twice that of the signal of part (b), that is, $f_H = 250$ Hz. Thus, the Nyquist rate is 500 Hz, and the Nyquist interval is $\dfrac{1}{500} = 2\text{ms}$.

12.6   BAND-PASS SAMPLING

The band-pass sampling theorem states that if a band-pass signal $g(t)$ has a spectrum of bandwidth B and an upper-frequency limit f_u, then $g(t)$ can be recovered from $g_s(t)$ by band-pass filtering if $f_s = 2f_u/k$, where k is the largest integer not exceeding f_u/B. All higher sampling rates are not necessarily usable unless they exceed $f_s = 2f_u$.

Example 12.3

Consider the band-pass signal $x(t)$ with the spectrum shown below (Figure 12.7):
    Check the band-pass sampling theorem by sketching the spectrum of the ideally sampled signal $x_s(t)$ when $f_s = 25$ kHz.

SOLUTION

From the figure above we have, $f_u = 30$ kHz and $B = 10$ kHz. Then $\dfrac{f_u}{B} = 3$ and $k = 2$. Hence, we have $f_s = \dfrac{2f_u}{k} = 30\,\mathrm{kHz}$ (Figure 12.8).

For $f_s = 30$ kHz, it can be seen that $g(t)$ is recovered from the sampled signal by using a band-pass filter. $g(t)$ can be recovered by using a low-pass filter with a cut-off frequency $f_c = 30$ kHz.

12.7   QUANTIZATION

The objective of the quantization process is to represent each sample by a fixed number of bits.

For example, if the amplitude of signal resulting from the sampling process ranges between (−1V and +1V), there is an infinite value of the voltage between (−1 and +1). For instance, one value can be −0.27689V. To assign a different binary sequence to each voltage value, we would have to construct a code of infinite length. Therefore, we can take a limited number of voltage values between (−1V and +1V) to represent the original signal, and these values must be discrete.

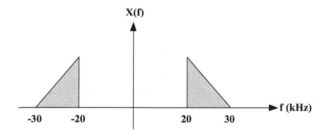

FIGURE 12.7   Signal of Example 12.2.

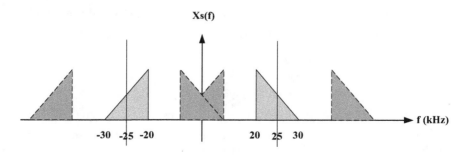

FIGURE 12.8   Sampled signal of Example 12.2.

Assume that the quantization steps were in 0.1V increments, and the voltage measurement for one sample is 0.58V. That would have to be rounded off to 0.6V, the nearest discrete value. Note that there is a 0.02V error, the difference between 0.58V and 0.6V. The error from the true value to the quantized value is called *quantization distortion* (Figure 12.9).

The higher the quantization level, the better quality of the system will deliver. However, increasing the quantization level has two major costs:

1) The cost of designing a system that needs a sizable binary code.
2) The time it takes to process this large number of quantizing steps by the encoder.

Therefore, a considerable number of quantizing levels may induce unwanted delays in the system.

12.8   UNIFORM AND NON-UNIFORM QUANTIZATION

From the above discussion, it is clear that the quantization noise depends on the step size. When the steps have a uniform size, the quantization is called *uniform quantization.*

For uniform quantization, the quantization noise is the same for all signal magnitudes. Therefore, with uniform quantization, the signal to noise ratio (SNR) is worse for low-level signals than for high-level signals.

Non-uniform quantization can provide fine quantization of the weak signal and coarse quantization of the strong signal. Thus in the case of non-uniform quantization, quantization noise can be made proportional to signal size. The effect is to improve the overall SNR by reducing the noise for the predominant weak signals, at the expense of an increase in noise for the rarely occurring strong signals.

Example 12.4

For the spectrum of the signal shown in Figure12.10, find the output signal if $f_m = 1$ kHz (Figure 12.10).

    a) $f_s = 3$ kHz
    b) $f_s = 4$ kHz
    c) $f_s = 6$ kHz

SOLUTION

    a) $f_s = 3$ kHz

$$f_s < 2f_m$$

Under-sampling (aliasing) (Figure 12.11).

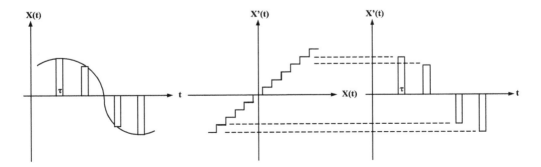

FIGURE 12.9   Concept of quantization.

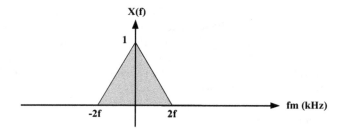

FIGURE 12.10   Signal of Example 12.4.

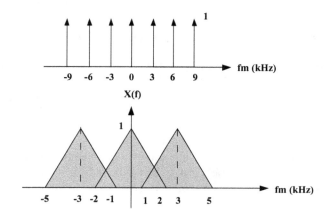

FIGURE 12.11   Sampled signal of Example 12.4 a.

b) $f_s = 4$ kHz

$$f_s = 2f_m = 2*2 = 4\,\text{kHz}$$

Low bandwidth (Figure 12.12).

c) $f_s = 6$ kHz

$$f_s > 2f_m$$

High bandwidth (Figure 12.13).

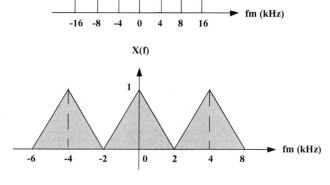

FIGURE 12.12   Sampled signal of Example 12.4 b.

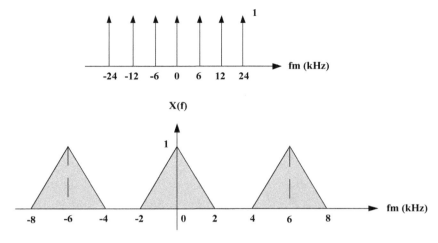

FIGURE 12.13   Sampled signal of Example 12.4 c.

Example 12.5

Sketch the following signals in discrete-time with $f_s = 4$ Hz and $N = 4$

a) $x(t) = 3\cos(2\pi t)$

b) $x(t) = 5\cos(200\pi t) + 2\cos(2\pi t)$

SOLUTION

$$t = nT_s, f_s = 4 = \frac{1}{T_s} = \frac{1}{4} = 0.25\,\text{sec}$$

a) $x(t) = 3\cos(2\pi t)$

$$x(nT_s) = 3\cos(2\pi nT_s) = 3\cos(2\pi \times 0.25 \times n)$$

$$x(0) = 3, x(1) = 0, x(2) = -3, x(3) = 0$$

b) $x(t) = 5\cos(200\pi t) + 2\cos(2\pi t)$

$$x(nT_s) = 5\cos(200\pi nT_s) + 2\cos(2\pi nT_s)$$

$$x(nT_s) = 5\cos(200\pi \times 0.25 \times n) + 2\cos(2\pi \times 0.25 \times n)$$

$$x(0) = 2, x(1) = 0, x(2) = -2, x(3) = 0$$

Example 12.6

The period of a periodic discrete-time function is 0.125 milliseconds, and it is sampled at 1,024 equally-spaced points. It assumed that with this number of samples, the sampling theorem is satisfied, and thus there will be no aliasing.

a) Compute the interval between frequency components in KHz

b) Compute the period of the frequency spectrum in KHz

c) Compute the sampling frequency fs

d) Compute the Nyquist frequency

SOLUTION

a) Intervals between samples and periods in discrete-time and frequency domains $T_t = 0.125$ milliseconds and $N = 1{,}024$ points. Therefore, the time between successive time components are

$$t_t = \frac{T_t}{N} = \frac{0.125 \times 10^{-3}}{1024} = 0.122\,\mu\text{sec}$$

b) the period Tf of the frequency spectrum is

$$f_t = \frac{1}{T_t} = \frac{1}{0.122 \times 10^{-3}} = 8.192\,\text{MHz}$$

c) the sampling frequency f_s is

$$f_s \geq 2f_t \geq 2 \times 8.192\,\text{MHz}$$

d) the Nyquist frequency must be equal or less than half the sampling frequency that is

$$f_s = 2f_t = 2 \times 8.192\,\text{MHz} = 16.384\,\text{MHz}$$

Example 12.7

Given an analog signal:

$$x(t) = 5\cos(2\pi \cdot 2000t) + 3\cos(2\pi \cdot 3000t), \quad \text{for } t \geq 0,$$

sampled at a rate of 8,000 Hz.

a. What is the Nyquist rate for this signal?
b. Sketch the spectrum of the sampled signal up to 20 kHz (Figure 12.14).
c. Sketch the recovered analog signal spectrum if an ideal low-pass filter with a cut-off frequency of 4 kHz is used to filter the sampled signal to recover the original signal (Figure 12.15).

SOLUTION

(a) For the given signal $f_{max} = 3{,}000$ Hz. According to the sampling theorem, the Nyquist rate/limit is $F_N = 2f_{max} = 6{,}000$ Hz. To avoid aliasing noise, the given signal should be sampled at a rate higher than F_N.

(b) Using Euler's formula, we get

$$x(t) = \frac{3}{2}e^{-j2\pi \cdot 3000t} + \frac{5}{2}e^{-j2\pi \cdot 2000t} + \frac{5}{2}e^{j2\pi \cdot 2000t} + \frac{3}{2}e^{j2\pi \cdot 3000t}$$

The spectrum of the sampled signal is sketched below:

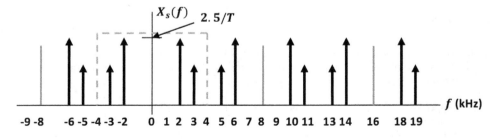

FIGURE 12.14 Sampled signal of Example 12.7 before filtering.

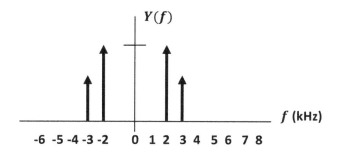

FIGURE 12.15 Sampled signal of Example 12.7.

(c) Based on the spectrum in part (b), the sampling theorem condition is satisfied. Hence, we can recover the original spectrum using a reconstruction low-pass filter with a cut-off frequency of 4 kHz. The recovered spectrum is sketched below:

12.9 AUDIO SAMPLING

Digital audio sampling is one of the most important applications of sampling; it is used in pulse-code modulation and digital signals for sound reproduction. This includes analog-to-digital conversion (ADC), digital-to-analog conversion (DAC), storage, and transmission. In effect, the system commonly referred to as digitalis is, in fact, a discrete-time, discrete-level analog of a previous electrical analog. While modern systems can be quite subtle in their methods, the primary usefulness of a digital system is the ability to store, retrieve, and transmit signals without any loss of quality.

12.10 SAMPLING RATE

A commonly seen unit of sampling rate is Hz, which stands for Hertz and means "samples per second." As an example, 48 kHz is 48,000 samples per second.

When it is necessary to capture audio covering the entire 20–20,000 Hz range of human hearing, such as when recording music or many types of acoustic events, audio waveforms are typically sampled at 44.1 kHz (CD), 48 kHz, 88.2 kHz, or 96 kHz; the approximately double-rate requirement is a consequence of the Nyquist theorem. Sampling rates higher than about 50 kHz to 60 kHz cannot supply more usable information for human listeners. Early professional audio equipment manufacturers chose sampling rates in the region of 40 to 50 kHz for this reason.

The Audio Engineering Society recommends a 48 kHz sampling rate for most applications but gives recognition to 44.1 kHz for Compact Discs (CD) and other consumer uses, 32 kHz for transmission-related applications, and 96 kHz for higher bandwidth or relaxed anti-aliasing filtering.

A more complete list of common audio sample rates is given in Table 12.1.

PROBLEMS

12.1 Find the Nyquist rate and the Nyquist interval for each of the following signals:

(a) $f(t) = 10\cos(2000\pi t)\cos(4000\pi t)$

(b) $f(t) = \dfrac{\sin(100\pi t)}{\pi t}$

(c) $f(t) = \left(\dfrac{\sin(100\pi t)}{\pi t}\right)^2$

TABLE 12.1
Common Audio Sample Rates

Sampling rate	Use
8,000 Hz	Telephone and encrypted walkie-talkie, wireless intercom and wireless microphone transmission; adequate for human speech but without sibilance.
11,025 Hz	Audio CDs; PCM, audio analysis.
16,000 Hz	Most modern VoIP and VVoIP communication products.
22,050 Hz	Audio CDs; PCM, MPEG audio and for audio analysis of low-frequency energy.
32,000 Hz	Digital video. video tapes, DAT (LP mode), NICAM digital audio, wireless microphones. digitizing FM radio.
37,800 Hz	CD-XA audio.
44,056 Hz	Used by digital audio locked.
44,100 Hz	Audio CD, audio (VCD, SVCD, MP3).
47,250 Hz	PCM sound recorder.
48,000 Hz	The standard audio sampling rate used by professional digital video equipment such as tape recorders, video servers, vision mixers, and so on. Mixing consoles and digital recording devices.
50,000 Hz	First commercial digital audio recorders from the late 70s from 3M and Sound stream.
50,400 Hz	The sampling rate used by the Mitsubishi X-80 digital audio recorder.
64,000 Hz	Uncommonly used, but supported by some hardware and software.
88,200 Hz	Some professional CD recording equipment.
96,000 Hz	DVD-Audio, some LPCM DVD tracks, audio tracks, HD DVD.
176,400 Hz	HDCD recorders and production.
192,000 Hz	DVD-Audio, DVD tracks, audio tracks, and HD DVD tracks,
352,800 Hz	Used for recording and editing Super Audio CDs.
2,822,400 Hz	Direct Stream Digital.
5,644,800 Hz	Some professional DSD recorders.
11,289,600 Hz	Uncommon professional DSD recorders.
22,579,200 Hz	Experimental DSD recorders.

12.2   The period of a periodic discrete-time function is 0.25 milliseconds, and it is sampled at 1,024 equally-spaced points. It is assumed that with this number of samples, the sampling theorem is satisfied, and thus there will be no aliasing.
   a. Compute the period of the frequency spectrum in KHz.
   b. Compute the interval between frequency components in KHz.
   c. Compute the sampling frequency fs.
   d. Compute the Nyquist frequency.

12.3   Given an analog signal:

$$x(t) = 5\cos(2\pi \cdot 1000t) + 3\cos(2\pi \cdot 2000t), \quad \text{for } t \geq 0,$$

sampled at a rate of 8,000 Hz.
   a) What is the Nyquist rate for this signal?
   b) Sketch the spectrum of the sampled signal up to 20 kHz.
   c) Sketch the recovered analog signal spectrum if an ideal low-pass filter with a cut-off frequency of 4 kHz is used to filter the sampled signal to recover the original signal.

12.4   Given an analog signal:

$$x(t) = 5\cos(2\pi \cdot 2500t) + 3\cos(2\pi \cdot 2000t), \quad \text{for } t \geq 0,$$

sampled at a rate of 8,000 Hz.

a. What is the Nyquist rate for this signal?
b. Sketch the spectrum of the sampled signal up to 25 kHz.
c. Sketch the recovered analog signal spectrum if an ideal low-pass filter with a cut-off frequency of 4 kHz is used to filter the sampled signal to recover the original signal.

12.5 Given an analog signal:

$$x(t) = 5\cos(2\pi \cdot 1500t), \quad \text{for } t \geq 0,$$

sampled at a rate of 8,000 Hz,
a. Sketch the spectrum of the original signal.
b. Sketch the spectrum of the sampled signal from 0 kHz to 20 kHz.

12.6 Given an analog signal:

$$x(t) = 5\cos(2\pi \cdot 2500t) + 2\cos(2\pi \cdot 3200t), \quad \text{for } t \geq 0,$$

sampled at a rate of 8,000 Hz.
a. Sketch the spectrum of the sampled signal up to 20 kHz.
b. Sketch the recovered analog signal spectrum if an ideal low-pass filter with a cut-off frequency of 4 kHz is used to filter the sampled signal to recover the original signal.

12.7 Given an analog signal:

$$x(t) = 5\cos(2\pi \cdot 2500t) + 2\cos(2\pi \cdot 4500t), \quad \text{for } t \geq 0,$$

sampled at a rate of 8,000 Hz.
a. Sketch the spectrum of the sampled signal up to 20 kHz.
b. Draw the recovered analog signal spectrum if an ideal low-pass filter with a cut-off frequency of 4 kHz is used to filter the sampled signal to recover the original signal.
c. Determine the frequency / frequencies of aliasing noise.

12.8 Given an analog signal:

$$x(t) = 10\cos(2\pi \cdot 5500t) + 5\sin(2\pi \cdot 7500t), \quad \text{for } t \geq 0,$$

sampled at a rate of 8,000 Hz.
a. Sketch the spectrum of the sampled signal up to 20 kHz.
b. Draw the recovered analog signal spectrum if an ideal low-pass filter with a cut-off frequency of 4 kHz is used to filter the sampled signal to recover the original signal.
c. Determine the frequency/frequencies of aliasing noise.

12.9 Consider the analog signal:

$$x(t) = 3\cos(100\pi t)$$

a. Determine the minimum sampling rate required to avoid aliasing.
b. Suppose that the signal is sampled at a rate of 200 Hz. What is the discrete-time signal obtained after sampling?
c. Suppose that the signal is sampled at the rate of $f_s = 75$ Hz. What is the discrete-time signal obtained after sampling?
d. What is the frequency $\dfrac{0 < F < f_s}{2}$ of a sinusoid signal that yields samples identical to those obtained in part (c)?

12.10 Given the analog signal:

$$x(t) = 3\cos(50\pi t) + 10\sin(300\pi t) - \cos(100\pi t), \quad \text{for } t \geq 0,$$

What is the Nyquist rate for this signal?

12.11 Consider the analog signal:

$$x(t) = 3\cos(2000\pi t) + 5\sin(6000\pi t) + 10\cos(12000\pi t), \quad \text{for } t \geq 0,$$

a. What is the Nyquist rate for this signal?
b. Assume now that we sample this signal using a sampling rate of 5,000 samples/sec. What is the discrete-time signal obtained after sampling?

12.12 If the analog signal to be quantized is a sinusoidal waveform, that is,

$$x(t) = 9.5\sin(2000 \times \pi t)$$

and if the bipolar quantizer uses 6 bits, determine
a. Number of quantization levels.
b. The quantization step size or resolution, Δ, assuming that the signal range is from -10 to 10 V.
c. The signal power to quantization power ration.

12.13 A digital communication link carries binary-coded words representing samples of an input signal

$$x_a(t) = 2\cos(600\pi t) + 2\cos(1800\pi t)$$

The link is operated at 10,000 bits/s, and each input sample quantized into 1,024 different voltage levels.
a. What is the sampling frequency and the folding frequency?
b. What is the Nyquist rate for the signal $x_a(t)$?
c. What are the frequencies in the resulting discrete-time signal $x[n]$?
d. What is the resolution Δ?

13 Digital Filters

Digital Filters are intended to pass signal components of certain frequencies without distortion.

To pass the signal, the frequency response should be equal to the signal's frequencies (pass -band). The frequency response should be equal to zero to block the signal. (stopband). This chapter will discuss the types of filters according to the pass signal and will discuss the Finite Impulse Response Digital Filter and Infinite impulse response digital filter.

13.1   TYPES OF FILTERS

There are four types of filter according to the pass frequencies.

13.1.1   LOW-PASS FILTERS

Low-pass filters are designed to pass low frequencies, from zero to a certain cut-off frequency and to block high frequencies.

13.1.2   HIGH-PASS FILTERS

High-pass filters are designed to pass high frequencies, from a certain cut-off frequency and to block low frequencies.

13.1.3   BAND-PASS FILTERS

Band-pass filters are designed to pass a certain frequency range, which does not include zero and to block other frequencies.

13.1.4   BAND-STOP FILTERS

Band-stop filters are designed to block a certain frequency range, which does not include zero and to pass other frequencies.

Figure 13.1 shows the types of these filters.

Example 13.1

Plot the magnitude frequency response for the low-pass filter given by the transfer function

$$H(z) = \frac{z}{(z - 0.25)}$$

SOLUTION

```
clear all
close all
clc
[h w] = freqz([1], [1 -0.25], 1024);
phi = 180*unwrap(angle(h))/pi;
figure
  plot(w, abs(h));
  grid on;
xlabel('Frequency (radians)');
ylabel('Magnitude');
```

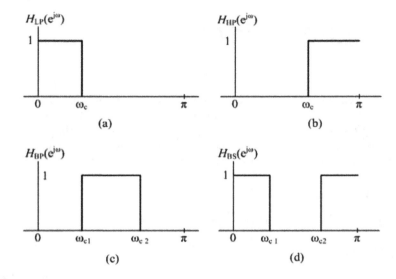

FIGURE 13.1   Types of filters.

Figure 13.2 shows a LPF Response.

Example 13.2

Plot the magnitude frequency response for the high-pass filter given by the transfer function

$$H(z) = 1 - 1.5 z^{-1}$$

SOLUTION

```
clear all
close all
```

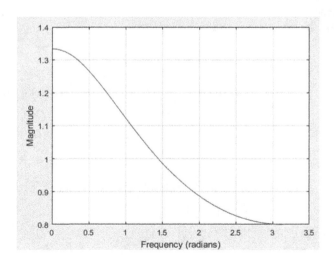

FIGURE 13.2   LPF response for Example 13.1.

```
clc
[h w] = freqz([1 -1.5], [1], 1024);
phi = 180*unwrap(angle(h))/pi;
figure
plot(w, abs(h));
grid on;
xlabel('Frequency (radians)');
ylabel('Magnitude');
```

Figure 13.3 shows a HPF Response.

Example 13.3

Plot the magnitude frequency response for the band-pass filter given by the transfer function

$$H(z) = \frac{0.5\,z^2 - 0.5}{z^2 - 0.25\,z + 0.25}$$

SOLUTION

```
clear all
close all
clc
[h w] = freqz([0.5 0 -0.5], [1 -0.25 0.25], 1024);
phi = 180*unwrap(angle(h))/pi;
figure
  plot(w, abs(h));
  grid on;
xlabel('Frequency (radians)');
ylabel('Magnitude');
```

Figure 13.4 shows a response of BPF.

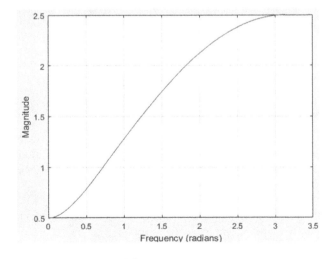

FIGURE 13.3 HPF response for Example 13.2.

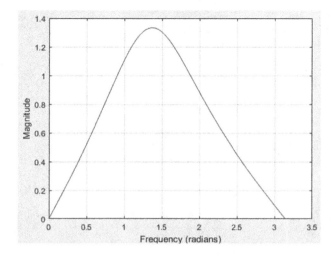

FIGURE 13.4 BPF response for Example 13.3.

Example 13.4

Plot the magnitude frequency response for the stop-pass filter given by the transfer function

$$H(z) = \frac{1 - 0.9\,z^{-1} + 0.81z^{-2}}{1 - 0.6\,z^{-1} + 0.36\,z^{-2}}$$

SOLUTION

```
clear all
close all
clc
[h w] = freqz([1 -0.9 0.81], [1 -0.6 0.16], 1024);
phi = 180*unwrap(angle(h))/pi;
figure
  plot(w, abs(h));
  grid on;
xlabel('Frequency (radians)');
ylabel('Magnitude');
```

Figure 13.5 shows a SPF Response.

13.2 INFINITE-IMPULSE-RESPONSE (IIR) DIGITAL FILTER

Digital IIR filters are derived from their analog counterparts. There are several common types of analog filters: Butterworth which has maximally flat pass-bands in filters of the same order, Chebyshev which have a ripple in the pass-band, and elliptic filters which are equi-ripple in both the pass-band and the stop-band.

- Our strategy will be to design the filter in the analog domain and then transform the filter to the digital domain.
- We can derive this transformation by recalling the relationship between the Laplace transform and the z-transform:
- This transformation is known as the *bilinear transform.*

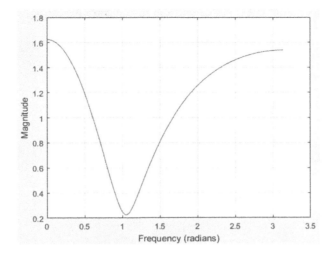

FIGURE 13.5 SPF response for Example 13.4.

13.2.1 DESIGN OF FILTERS USING BILINEAR TRANSFORMATION

A method to map the left half of the s-plane to the inside unit circle of the z-plane.

Mapping of the s-plane to the z-plane

- A point on the $j\Omega$-axis in the s-plane mapped onto a point on the unit circle in the z-plane.
- A point in the left-half of the s-plane with $\sigma < 0$ mapped inside of the unit circle in the z-plane.
- A point in the right-half of the s-plane with $\sigma > 0$ mapped outside of the unit circle in the z-plane.

Figure 13.6 shows the s- and the z-plane.

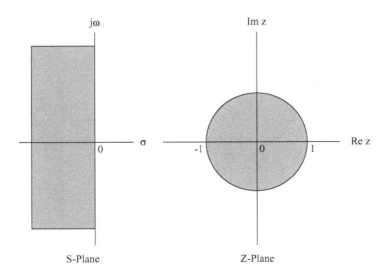

FIGURE 13.6 The s- and z-plane.

- The design of the digital IIR Filter, $H(z)$, from the analog filter, $H_a(s)$, requires a mapping of the s-plane to the z-plane.
- The point in the left-half of the s-plane should map to points inside the unit circle to preserve the stability of the analog filter.
- Bilinear transformation is given by

$$s = \frac{2}{T}\left(\frac{z-1}{z+1}\right)$$ (13.1)

- Maps a single point in the s-plane to a unique point in the z-plane.
  - The relationship

$$H(z) = H(s)\Big|_{z \to \frac{2}{T}\left(\frac{z-1}{z+1}\right)}$$ (13.2)

- Frequency wrapping,

$$\Omega = \frac{2}{T}\tan\left(\frac{\omega}{2}\right)$$ (13.3)

- Choose $T = 2$ to simplify the design procedure.
- So, the parameter T does not affect $H(z)$.

$$z = \frac{1+s}{1-s}$$ (13.4)

Example 13.5

Convert the following analog first-order low-pass filter into a digital filter given as the analog transfer function

$$H_{LP}(s) = \frac{\Omega_c}{s + \Omega_c}$$

SOLUTION

Applying bilinear transformation,

$$G_{LP}(z) = \frac{\Omega_c}{s + \Omega_c}\Bigg|_{s=\frac{2}{T}\left(\frac{1-z^{-1}}{1+z^{-1}}\right)} = \frac{\dfrac{\Omega_c T}{2}(1+z^{-1})}{(1-z^{-1}) + \dfrac{\Omega_c T}{2}(1+z^{-1})}$$

$$= \frac{1-a}{2}\left(\frac{1+z^{-1}}{1-az^{-1}}\right) \quad \text{where} \quad a = \frac{1 - \tan(\omega_c T / 2)}{1 + \tan(\omega_c T / 2)}$$

Example 13.6

Convert the following analog first-order high-pass filter into digital filter given the analog transfer function:

$$H_{HP}(s) = \frac{s}{s + \Omega_c}$$

SOLUTION

Applying bilinear transformation,

$$G_{HP}(z) = \frac{s_c}{s + \Omega_c}\bigg|_{s=\frac{2}{T}\left(\frac{1-z^{-1}}{1+z^{-1}}\right)} = \frac{1+a}{2}\left(\frac{1-z^{-1}}{1-az^{-1}}\right)$$

13.2.2 Infinite-Impulse Response Filtering

This section introduces IIR filters. In this chapter, we will study some commonly used IIR-filtering structures, design processes, and implementations on both fixed-point and floating-point DSP processors. Also, important practical issues such as quantization effects in different IIR-filter structures, fixed-point implementations, and scaling problems are examined using FDATool. An application that uses IIR filters for DTMF generation and detection is presented at the end of the chapter.

FIR-filter characteristics were introduced in this chapter and briefly compared with IIR filters. In general, IIR filters have the following features:

1. Nonlinear phase: IIR filters have a nonlinear-phase response over the frequency of interest. Therefore, group delay varies at different frequencies and results in phase distortion.
2. Stability issue: IIR filters are not always stable due to their recursive realization. Therefore, a careful *d* sign approach is needed to ensure that all of the poles of an IIR filter lie inside the unit circle to guarantee a stable filter, especially for fixed-point implementations as well as the inputs. Due to the nature of output feedback, there is an inherent one sample delay in the feedback section. The IIR filter performs two inner (or dot) products of vectors, one is for the feedforward section between vectors *b* and *x*, and the other for the feedback section between vectors *a* and *y*. The design of the IIR filter determines two sets of coefficients, $\{bj, i = 0,1, \ldots, L-1\}$ and $\{am, m = 1,2, \ldots, M-1\}$, to meet a given specification. Also, the IIR filter is recursive in computation, which results in an infinite-impulse-response. Therefore, the IIR filter must be designed with special care to prevent any growing or oscillation of the impulse response that can lead to an unstable filter.

By taking the *z*-transform using the time-shift property, we have

$$Y(z) = \left(b_0 + b_1 z^{-1} + \cdots + b_{L-1}Z^{-L+1}\right)X(Z) - \left(a_1 z^{-1} + a_2 z^{-2} + \cdots + a_{M-1}z^{-M+1}\right)Y(z) \tag{13.5}$$

$$= x(z)\sum_{i=0}^{l-1} b_i z^{-i} - y(z)\sum_{m=1}^{m-1} a_m z^{-m}$$

By rearranging the terms, we obtain the transfer function of an IIR filter expressed as

$$x(z) = \frac{y(z)}{x(z)}$$

$$= \frac{\displaystyle\sum_{i=0}^{l-1} b_i z^{-1}}{1 + \displaystyle\sum_{m+1}^{m-1} a_m z^{-m}} = \frac{b_0 + b_1 z^{-1} + \cdots + b_{L-1}Z^{-L+1}}{a_0 + a_1 z^{-1} + \cdots a_{m-1}z^{-m+1}} \tag{13.6}$$

The IIR filter is expressed as the ratio of polynomials in $z-l$ with $(L-1)$ zeros and $(M-1)$ poles. The roots of the numerator and the denominator polynomials determine zeros and poles, respectively. To ensure a stable filter, all of the poles must be placed inside the unit circle in the *z*-plane. The IIR filter provides greater flexibility in the filter design since both the poles and zeros contribute to the frequency response.

13.2.3 Filter Characteristics

The coefficient vector of the FIR filter is identical to the impulse response of the filter. However, the impulse response of the IIR filter is not similar to the coefficient vector. The length of the impulse response for the IIR filter can be infinity; thus, the filter is called the infinite-impulse-response filter.

The transfer function is given in Equation (13.6) and can be evaluated on the unit circle to obtain the frequency response expressed as

$$H(\omega) = H(z)_{z=e^{jw}} = \frac{\sum_{i=0}^{L-1} b_i e^{-j\omega i}}{1 + \sum_{i=0}^{m-1} a_m e^{-j\omega m}} \tag{13.7}$$

The frequency response is a function of the continuous-frequency variable and is a periodic function with a period equal to 2. Therefore, we only need to evaluate the frequency response from 0 to 2π or from $-\pi$ to π. The DFT can be applied to evaluate the frequency response $H(\omega)$ at equally-spaced frequency points $\omega k = 2\pi k/N$, $k = 0,1, \ldots , N-1$.

The magnitude response $|H(\omega)|$ and the phase response $\phi(\omega)$ are periodic functions with a period of 2π. If the IIR filter has real-valued coefficients, the real part of the frequency response is an even function, and the imaginary part is an odd function. As such, the magnitude and phase responses are even and odd functions, respectively, expressed as

$$|X(\omega)| = |X(-\omega)| \quad \text{and} \quad \phi(\omega) = (-\phi). \tag{13.8}$$

Therefore, we only need to compute the magnitude and phase responses from 0 to π for IIR filters with real coefficients.

The FIR filter with symmetric coefficients has a linear-phase response. The IIR filter, however, is not able to guarantee a linear-phase response, which also implies that the IIR filter is not able to produce a constant phase (or group) delay.

Given an IIR filter with coefficient vector $b = [1\ 1.5]$ and $a = [1\ -0.25\ 0.6]$, the transfer function of the filter given as

$$H(z) = \frac{1 + 0.5\,z}{1 - 0.25\,z^{-1} + 0.6\,z^{-2}}$$

This function shows that the impulse response is no longer related to the coefficient vectors b and a.

We can use the MATLAB functions impz, freqz, grpdelay, and z-plane to analyze the preceding transfer function as follows:

```
clc;
clear all;
b =[1  1.5]
a = [1 -.25 0.6]
figure
impz(b,a); pause;
zplane(b,a); pause;
freqz(b,a); pause;
grpdelay(b,a); pause;
```

The impulse response, pole-zero diagram, and magnitude and phase responses are shown in Figure 13.7. From these results, we observe that the impulse response of a stable IIR filter converges (decays) to a minimum value as time increases. The pole-zero plot shows both the pole and zero locations. For a stable IIR filter, all poles must lie inside the unit circle. From the magnitude

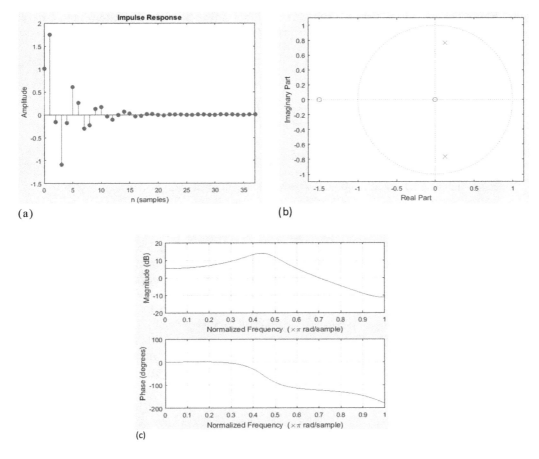

FIGURE 13.7 Characteristics of an IIR filter: (a) Impulse response, (b) pole-zero diagram, and (c) magnitude and phase responses.

response shown in Figure 13.7c, the filter is a low-pass filter. However, the phase is nonlinear within the pass-band; thus, the group delay is no longer a constant over the pass-band.

13.3 FINITE IMPULSE RESPONSE (FIR) DIGITAL FILTER

FIR

The *finite impulse response* filter is a filter whose impulse response is of a *finite* duration because it settles to zero in finite time. This contrasts with infinite-impulse-response filters, which may have internal feedback and may continue to respond indefinitely (usually decaying).

A FIR digital filter has a finite number of non-zero coefficients of its impulse response:

$$h(n) = 0 \quad \text{for } |n| > M$$

The mathematical model of a causal FIR digital filter:

$$y(n) = \sum_{k=0}^{M-1} h(k)x(n-k) \tag{13.9}$$

Digital FIR filters cannot be derived from analog filters since causal analog filters cannot have a finite impulse response. In many digital signal processing applications, FIR filters are preferred over their IIR counterparts.

13.3.1 THE ADVANTAGES OF FIR FILTERS

- FIR filters have a linear phase. Linear-phase filters are important for applications where frequency dispersion due to nonlinear phase is harmful (e.g., speech processing).
- FIR filters are stable.
- Excellent design methods are available for various kinds of FIR filters.
- Infection finite-precision errors: FIR filters are less sensitive to finite-precision errors effects such as coefficient quantization errors and round-off noise.

13.3.2 FIR SPECIFICATIONS

A filter specification is usually based on the desired magnitude response. For example, the specification for a low-pass filter illustrated in Figure 13.8. The pass-band is defined as the frequency range over which the input signal is passed with approximate unit gain. Thus, the pass-band is defined by $0 \le \omega \le \omega 1$, where wp is called the pass-band-cut-off (edge) frequency. The width of the pass-band is usually called the bandwidth of the filter. For a low-pass filter, the bandwidth is ω_1. The input-signal components that lie within the stop-band are attenuated to a level that effectively eliminates them from the output signal. The stop-band is defined by $\omega_1 \le \omega \le \omega_2 \le \pi$, where ω_2 is called the stop-band-edge frequency. The transition band is defined as $\omega_1 < \omega < \omega_2$, which is between the pass-band and the stop-band. In this region, the filter magnitude response is typically,

$$A_1 = 20\text{Log}_{10}\left(1+\delta_1\right) \ db \tag{13.10}$$

and

$$A_2 = -20\text{Log}_{10}(\delta_2) \ db \tag{13.11}$$

For example, suppose that $\delta_1 = 0.1$ and $\delta_2 = 0.01$, we have $A_1 = 20\log_{10} (1.1) = 0.828$ dB and $A_2 = -20 \log_{10} (0.01) = 40$ dB.

There are two MATLAB functions (syntax given below), that can be used to perform this filtering process:

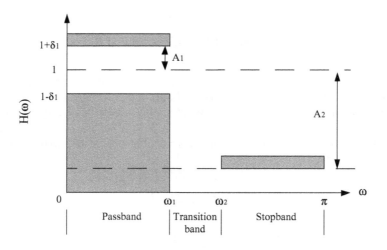

FIGURE 13.8 Low-pass filter specification.

$$Z_i = \text{filter}\left(B, A, Y_i, X_i\right)$$

$$y = \text{filter}\left(B, A, x, Z_i\right)$$

where B and A are vectors for the coefficients given as

$$A = \begin{bmatrix} 1\, a_1\, a_2 \dots a_N \end{bmatrix} \text{ and } B = \begin{bmatrix} b_0\, b_1\, b_2 \dots b_M \end{bmatrix}$$

Xi and *Yi* are the vectors containing the initial conditions. Also x,y is the input and system output vectors.

The function filtic is used to obtain the initial states required by the second function filter. The function filter is based on the *direct-form II realization* to implement a digital filter from its difference equation form.

The following MATLAB code illustrates how to solve by using filtic and filter MATLAB functions.

```
clear all;
b = [0 1 2];
a = [1 0 -2.5];
Xi = [-1 0];
Yi = [0 1];
Zi = filtic(b, a, Yi, Xi);
n = 0:5;
x = (0.95).^n;
y = filter(b, a, x, Zi)
y = 1.5000 -1.0000 6.7000 0.3025 19.4124 3.2855
```

13.3.3   GIBBS PHENOMENON AND DIFFERENT WINDOWING

Direct truncation of the impulse response leads to the well-known Gibbs phenomenon. It manifests itself as a fixed percentage overshoot and ripple before and after a discontinuity in the frequency response.

The effect of the windowing is that the ripple from the side lobes produces a ripple in the resulting frequency response. The windows may satisfy some reasonable optimality criterion:

a) The small width of the main portion of the frequency response of the window containing as much the total energy as possible.
b) Side lobes of the frequency response that decrease in energy rapidly as ω tends to π.

13.4   COMPARISON OF IIR AND FIR DIGITAL FILTERS

IIR type digital filters have the advantage of being economical in their use of delays, multipliers, and adders. They have the disadvantage of being sensitive to coefficient round-off inaccuracies and the effects of overflow in fixed-point arithmetic. These effects can lead to instability or serious distortion. Also, an IIR filter cannot be an exactly linear phase.

FIR filters may be realized by non-recursive structures which are simpler and more convenient for programming, especially on devices specifically designed for digital signal processing. These structures are always stable, and because there is no recursion, round-off, and overflow errors are easily controlled. A FIR filter can be a linear phase. The main disadvantage of FIR filters is that large orders can be required to perform fairly simple filtering tasks. Table 13.1 shows the important difference between using IIR or FIR.

TABLE 13.1

Difference between IIR and FIR

	IIR	FIR
Efficiency	More than FIR.	Less.
Design	Complex in design than FIR.	Easy to design.
Phase Linearity	Difficult to control, nonlinear.	Linear phase always possible.
Stability	Can be unstable because of feedback.	Always stable (no feedback).
Order	Less.	More.
History	Derived from analog filters.	No analog history.

PROBLEMS

13.1 Given the difference equation

$$y(n) = x(n-1) - 0.75y(n-1) + 0.125y(n-2)$$

a. Use the MATLAB functions **filter()** and **filtic()** to calculate the system response $y(n)$ for $n = 0,1,2,3,4$ with the input of $x(n) = (0.5)^n u(n)$ and initial conditions: $x(-1) = -1, y(-2) = 2$ and $y(-1) = 1$;

b. Use the MATLAB functions **filter()** and **filtic()** to calculate the system response $y(n)$ for $n = 0,1,2,3,4$ with the input of $x(n) = (0.5)^n u(n)$ and zero initial conditions: $x(-1) = 0, y(-2) = 0$ and $y(-1) = 0$.

13.2 Given the filter

$$H(z) = \frac{1 - z^{-1} + z^{-2}}{1 - 0.9z^{-1} + 0.81z^{-2}}$$

a. Plot the magnitude frequency response and phase response using MATLAB.

b. Specify the type of filtering.

c. Find the difference equation.

d. Perform filtering, that is, calculate $y(n)$ for the first 1,000 samples for each of the following inputs using MATLAB, assuming that all initial conditions are zeros and the sampling rate is 8,000Hz:

(1) $x(n) = \cos\left(\pi \cdot 10^3 \cdot n / 8000\right)$

(2) $x(n) = \cos\left(6\pi \cdot 10^3 \cdot n / 8000\right)$

(3) $x(n) = \cos\left(\frac{8}{3}\pi \cdot 10^3 \cdot n / 8000\right)$

e. Repeat part (d) using MATLAB function **filter()**.

13.3 Find $H(z)$ for the following difference equations

(a) $y[n] = 2x[n] - 3x[n-1] + 6x[n-4]$

(b) $y[n] = x[n-1] - y[n-1] - 0.5y[n-2]$

13.4 Show that passing any sequence $\{x[n]\}$ through a system with $H(z) = z^{-1}$ produces $\{x[n-1]\}$ i.e., all samples are delayed by one sampling interval.

13.5 Calculate the impulse response of the digital filter with

$$H(z) = \frac{1}{1 - 2z^{-1}}$$

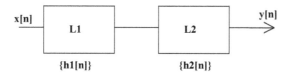

FIGURE 13.9 System for Problem 13.7.

13.6 Draw the signal-flow graph, for Example 13.5 and plot its poles and zeros.

13.7 If discrete-time LTI systems L_1 and L_2, with impulse responses $\{h_1[n]\}$ and $\{h_2[n]\}$, respectively, are serially cascaded, as shown in Figure 13.9, calculate the overall impulse response. Show that this will not be affected by interchanging L_1 and L_2.

13.8 Design a fourth-order band-pass IIR digital filter with lower and upper cut-off frequencies at 300 Hz and 3,400 Hz when $fS = 8$ kHz.

13.9 Design a fourth-order band-pass IIR digital filter with lower and upper cut-off frequencies at 2,000 Hz and 3,000 Hz when $fS = 8$ kHz.

13.10 What limits how good a notch filter we can implement on a fixed-point DSP processor? In theory, we can make a notch sharper and sharper by reducing the –3dB bandwidth and/ or increasing the order. What limits us in practice?
How sharp can a notch get in 16-bit fixed pt arithmetic?

13.11 What order of FIR filter would be required to implement a $\pi/4$ notch approximately as good as a second-order IIR $\pi/4$ notch with 3 dB bandwidth 0.2 radians/sample?

13.12 What order of FIR low-pass filter would be required to be approximately as good as the second-order IIR low-pass filter ($\pi/4$ cut-off) designed in these notes?

14 Implementation of IIR

There are many different structures for realizing digital IIR filters, and we need to understand and determine a suitable arrangement for a particular application. The factors to be considered include errors in quantizing filter coefficients, noise introduced by fixed-point arithmetic, propagation of quantization errors, computational load, memory usage, and programming consideration and flexibility. In this section, we examine the following commonly used IIR filter structures:

1. Direct form I
2. Direct form II
3. Cascade form
4. Parallel form
5. Transposed direct form I
6. Transposed direct form II

14.1 DIRECTION-FORM I REALIZATION

The transfer function of a digital filter is given by

$$H(z) = \frac{Y(z)}{X(z)} = \frac{b_0 + b_1 z^{-1} + \cdots + b_M z^{-M}}{1 + a_1 z^{-1} + \cdots + a_N z^{-N}} = \frac{B(z)}{A(z)} \tag{14.1}$$

This expression can also be written as

$$Y(z) = H(z)X(z) \tag{14.2}$$

or

$$Y(z) = \left(\frac{b_0 + b_1 z^{-1} + \cdots + b_M z^{-M}}{1 + a_1 z^{-1} + \cdots + a_N z^{-N}} \right) X(z) \tag{14.3}$$

Taking the inverse z-transform of the Equation (14.2) yields the difference equation

$$y(n) = b_0 x(n) + b_1 x(n-1) + \cdots + b_M x(n-M)$$

$$-a_1 y(n-1) - a_2 y(n-2) - \cdots - a_N y(n-N) \tag{14.4}$$

This difference equation can be realized by a direct-form I realization. We introduce the product and delay block diagram as shown in Figure 14.1.

The direct-form I realization of the difference Equation (14.3) is given in Figure 14.2.

The direct-form I realization of the second-order IIR filter ($N = M = 2$) is given by

$$y(n) = b_0 x(n) + b_1 x(n-1) + b_2 x(n-2) - a_1 y(n-1) - a_2 y(n-2) \tag{14.5}$$

As shown in Figure 14.3.

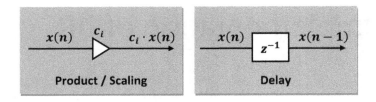

FIGURE 14.1   Product and delay block diagram.

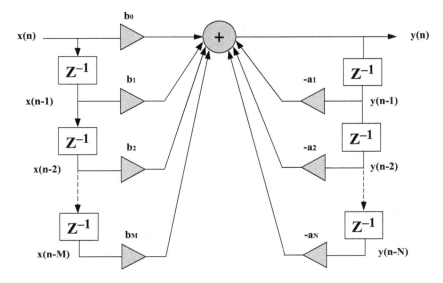

FIGURE 14.2   Direct-form I structure.

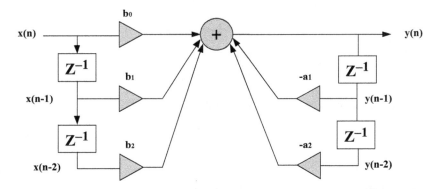

FIGURE 14.3   Second-order direct-form I structure.

14.2   DIRECTION-FORM II REALIZATION

Using the digital filter transfer function, we can write

$$Y(z) = H(z)X(z) = \frac{B(z)}{A(z)}X(z) = B(z)\left(\frac{X(z)}{A(z)}\right) \tag{14.6}$$

The expression can also be written as

$$Y(z) = \left(b_0 + b_1 z^{-1} + \cdots + b_M z^{-M}\right)\left(\frac{X(z)}{1 + a_1 z^{-1} + \cdots + a_N z^{-N}}\right) \tag{14.7}$$

By defining

$$W(z) = \frac{1}{1 + a_1 z^{-1} + \cdots + a_N z^{-N}} X(z) \tag{14.8}$$

We have

$$Y(z) = \left(b_0 + b_1 z^{-1} + \cdots + b_M z^{-M}\right) W(z) \tag{14.9}$$

The corresponding difference equations for Equations (14.2) and (14.3) are

$$w(n) = x(n) - a_1 w(n-1) - a_2 w(n-2) - \cdots - a_N w(n-N) \tag{14.10}$$

and

$$y(n) = b_0 w(n) + b_1 w(n-1) + \cdots + b_M w(n-M) \tag{14.11}$$

These difference equations can be implemented by a direct-form II realization as shown in Figure 14.4.

The direct-form II realization of the second-order IIR filter ($N = M = 2$) is given by the equations.

$$\boldsymbol{w(n) = x(n) - a_1 w(n-1) - a_2 w(n-2)} \tag{14.12}$$

$$\boldsymbol{y(n) = b_0 w(n) + b_1 w(n-1) + b_2 w(n-2)} \tag{14.13}$$

As shown in Figure 14.5.

14.3 CASCADE (SERIES) REALIZATION

The digital filter transfer function can be factorized and written as

$$\boldsymbol{H(z) = H_1(z) \cdot H_2(z) \cdots H_k(z)} \tag{14.14}$$

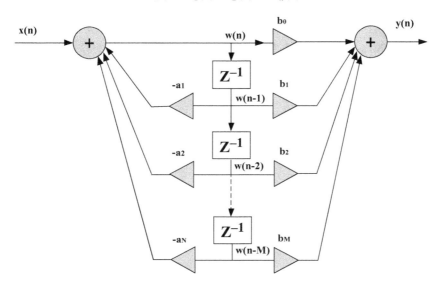

FIGURE 14.4 Direct-form II structure.

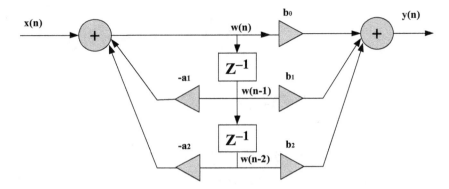

FIGURE 14.5   Second-order direct-form II structure.

where $H_k(z)$ is chosen as the first- or second-order transfer function (section), defined as,

$$H_k(z) = \left(\frac{b_{k0} + b_{k1}z^{-1}}{1 + a_{k1}z^{-1}} \right) \tag{14.15}$$

or

$$H_k(z) = \left(\frac{b_{k0} + b_{k1}z^{-1} + b_{k2}z^{-2}}{1 + a_{k1}z^{-1} + a_{k2}z^{-2}} \right) \tag{14.16}$$

The block diagram for the cascade (or series) realization is shown in Figure 14.6.

For the individual first or second-order transfer function (section), one can use either direct-form I or direct-form II realizations.

14.4   PARALLEL REALIZATION

The digital filter transfer function can be written as

$$H(z) = H_1(z) + H_2(z) + \cdots + H_k(z) \tag{14.17}$$

where $H_k(z)$ is chosen as the first- or second-order transfer function (section), defined as,

$$H_k(z) = \left(\frac{b_{k0}}{1 + a_{k1}z^{-1}} \right) \tag{14.18}$$

or

$$H_k(z) = \left(\frac{b_{k0} + b_{k1}z^{-1}}{1 + a_{k1}z^{-1} + a_{k2}z^{-2}} \right) \tag{14.19}$$

The block diagram for the parallel realization is shown in Figure 14.7.

Here again, for the individual first- or second-order transfer function (section), one can use either direct-form I or direct-form II realizations.

$$x(n) \longrightarrow \boxed{H_1(z)} \longrightarrow \boxed{H_2(z)} \longrightarrow \cdots \longrightarrow \boxed{H_K(z)} \xrightarrow{\;y(n)\;}$$

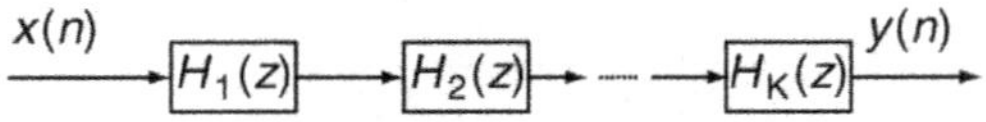

FIGURE 14.6   Cascade (or series) realization.

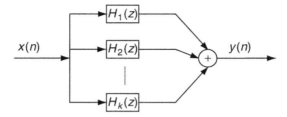

FIGURE 14.7   Parallel realization.

Example 14.1

Given a second-order transfer function

$$H(z) = \frac{0.5\left(1 - z^{-1}\right)}{\left(1 + 1.3z^{-1} + 0.36z^{-2}\right)}$$

Perform the filter realizations and write the difference equations using the following realizations

(1) Direct-form I and direct-form II
(2) Cascade form via the first-order sections
(3) Parallel form via the first-order sections

SOLUTION

Part (1): The transfer function in its "delay" form can be written as

$$H(z) = \frac{0.5 - 0.5z^{-2}}{1 + 1.3z^{-1} + 0.36z^{-2}}$$

Comparing it with the standard delay form of the transfer function

$$H(z) = \frac{b_0 + b_1 z^{-1} + \cdots + b_M z^{-M}}{1 + a_1 z^{-1} + \cdots + a_N z^{-N}}$$

We identify, that $M = 1$, $N = 2$, and

$$b_0 = 0.5, \quad b_1 = 0.0, \quad b_2 = -0.5$$

$$a_1 = 1.3, \quad a_2 = 0.36$$

The difference equation form for the direct-form I realization is given by

$$y(n) = 0.5\,x(n) - 0.5\,x(n-2) - 1.3\,y(n-1) - 0.36\,y(n-2)$$

The direct-form I realization for this filter shown in Figure 14.8.
   For the direct-form II realization, we write the difference equation as made up of the following two difference equations

$$w(n) = x(n) - 1.3w(n-1) - 0.36w(n-2)$$

$$y(n) = 0.5w(n) - 0.5w(n-2)$$

The direct-form II realization of the filter is shown below in Figure 14.9.
   Part (2): For the cascade realization, the transfer function is written in the product form. This is achieved by factorization of the numerator and the denominator polynomials. The given transfer function is

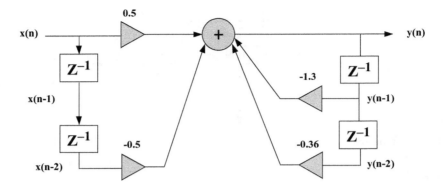

FIGURE 14.8 The direct-form I realization of Example 14.1.

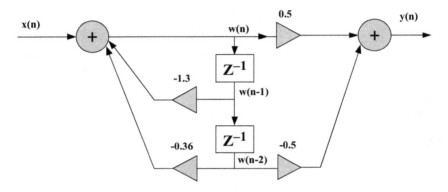

FIGURE 14.9 The direct-form II realization of Example 14.1.

$$H(z) = \frac{0.5\left(1 - z^{-2}\right)}{1 + 1.3z^{-1} + 0.36z^{-2}}$$

The numerator polynomial can be factorized as

$$B(z) = 0.5\left(1 - z^{-2}\right) = 0.5\left(1 - z^{-1}\right)\left(1 + z^{-1}\right)$$

The denominator polynomial can be factorized as

$$A(z) = 1 + 1.3z^{-1} + 0.36z^{-2} = 1 + 0.4z^{-1} + 0.9z^{-1} + 0.36z^{-2}$$

$$= 1\left(1 + 0.4z^{-1}\right) + 0.9z^{-1}\left(1 + 0.4z^{-1}\right)$$

$$= \left(1 + 0.4z^{-1}\right)\left(1 + 0.9z^{-1}\right)$$

Therefore, the transfer function can be written as

$$H(z) = \frac{0.5\left(1 - z^{-1}\right)\left(1 + z^{-1}\right)}{\left(1 + 0.4z^{-1}\right)\left(1 + 0.9z^{-1}\right)}$$

or

$$H(z) = \left(\frac{0.5 - 0.5z^{-1}}{1 + 0.4z^{-1}}\right)\left(\frac{1 + z^{-1}}{1 + 0.9z^{-1}}\right) = H_1(z) \cdot H_2(z)$$

Thus, in this case

$$H_1(z) = \frac{0.5 - 0.5z^{-1}}{1 + 0.4z^{-1}}$$

$$H_2(z) = \frac{1 + z^{-1}}{1 + 0.9z^{-1}}$$

$H_1(z)$ and $H_2(z)$ can both be realized in direct-form I or direct-form II. Overall, we get the cascaded realization with two sections. It should be noted that there could be other forms for $H_1(z)$ and $H_2(z)$; for example, we could have taken $H_1(z) = \dfrac{1 + z^{-1}}{1 + 0.4z^{-1}}$, $H_2(z) = \dfrac{0.5 - 0.5z^{-1}}{1 + 0.9z^{-1}}$, to yield the same $H(z)$. Using the former $H_1(z)$ and $H_2(z)$, and using direct-form II realizations for the two cascaded sections as shown in Figure 14.10, we get the following difference equations:

Part (1): $\left(H_1(z) = \dfrac{0.5 - 0.5z^{-1}}{1 + 0.4z^{-1}} \right)$

$$w_1(n) = x(n) - 0.4w(n-1)$$

$$y_1(n) = 0.5w_1(n) - 0.5w_1(n-1)$$

Part (2): $\left(H_2(z) = \dfrac{1 + z^{-1}}{1 + 0.9z^{-1}} \right)$

$$w_2(n) = y_1(n) - 0.9w_2(n-1)$$

$$y(n) = w_2(n) + w_2(n-1)$$

Part (3): For the parallel realization, the transfer function is written in partial fraction form. The given transfer function is

$$H(z) = \frac{0.\left(1 - z^{-2}\right)}{1 + 1.3z^{-1} + 0.36z^{-2}}$$

Multiplying the numerator and denominator with z^2 we get

$$H(z) = \frac{0.5\left(z^2 - 1\right)}{z^2 + 1.3z + 0.36}$$

The denominator polynomial can be factorized as

$$A(z) = z^2 + 1.3z + 0.36 = z^2 + 0.4z + 0.9z + 0.36$$

$$= 1(z + 0.4) + 0.9z(z + 0.4) = (z + 0.4)(z + 0.9)$$

Therefore, the transfer function can be written as

$$H(z) = \frac{0.5\left(z^2 - 1\right)}{(z + 0.4)(z + 0.9)}$$

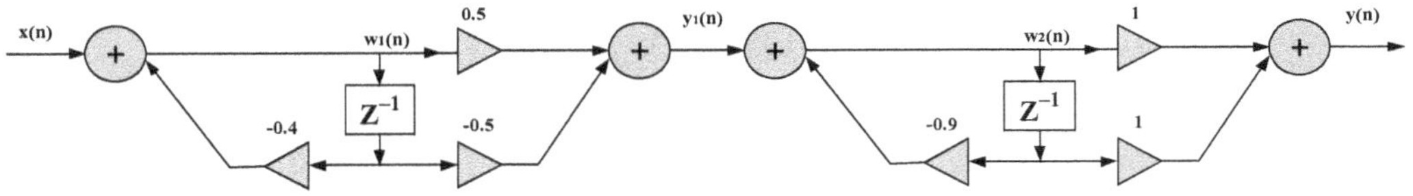

FIGURE 14.10 Cascaded realization Example 14.1.

or

$$\frac{H(z)}{z} = \frac{0.5(z^2 - 1)}{z(z+0.4)(z+0.9)}$$

Writing it into partial fractions form,

$$\frac{H(z)}{z} = \frac{A}{z} + \frac{B}{(z+0.4)} + \frac{C}{(z+0.9)}$$

where

$$A = z\left(\frac{0.5(z^2 - 1)}{z(z+0.4)(z+0.9)}\right)\Bigg|_{z=0} = \frac{0.5(z^2 - 1)}{(z+0.4)(z+0.9)}\Bigg|_{z=0} = -1.39$$

$$B = (z+0.4)\left(\frac{0.5(z^2 - 1)}{z(z+0.4)(z+0.9)}\right)\Bigg|_{z=-0.4} = \frac{0.5(z^2 - 1)}{z(z+0.9)}\Bigg|_{z=-0.4} = -2.1$$

$$C = (z+0.9)\left(\frac{0.5(z^2 - 1)}{z(z+0.4)(z+0.9)}\right)\Bigg|_{z=-0.9} = \frac{0.5(z^2 - 1)}{z(z+0.4)}\Bigg|_{z=-0.9} = 0.21$$

Therefore,

$$\frac{H(z)}{z} = \frac{-1.39}{z} + \frac{-2.1}{(z+0.4)} + \frac{0.21}{(z+0.9)}$$

or

$$H(z) = -1.39 - \frac{2.1z}{(z+0.4)} + \frac{0.21z}{(z+0.9)}$$

In delay form, it can be written as

$$H(z) = -1.39 - \frac{2.1}{(1+0.4z^{-1})} + \frac{0.21}{(1+0.9z^{-1})}$$

or we can write it as

$$H(z) = H_1(z) + H_2(z) + H_3(z)$$

This filter shows that there are three sections in the parallel realization. They can be individually realized using either direct-form I or direct-form II realizations. We use the direct-form II realization. The difference equations for each of the three parallel sections are

Section 1: $(H_1(z) = -1.39)$

$$y_1(n) = -1.39\, x(n)$$

Section 2: $\left(H_2(z) = -\dfrac{2.1}{(1+0.4z^{-1})}\right)$

$$w_2(n) = x(n) - 0.4w_2(n - 1)$$

$$y_2(n) = 2.1w_2(n)$$

Section 3: $\left(H_3(z) = \dfrac{0.21}{(1+0.9z^{-1})}\right)$

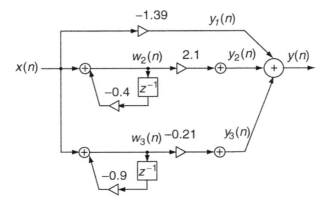

FIGURE 14.11 Parallel realization Example 14.1.

$$w_3(n) = x(n) - 0.9w_3(n-1)$$

$$y_3(n) = -0.21w_3(n)$$

The output is $y(n) = y_1(n) + y_2(n) + y_3(n)$ and the parallel realization is shown in Figure 14.11.

14.5 TRANSPOSED-DIRECT-FORM-I

The transposition theorem uses to convert the direct-form I realization into the transposed IIR filter shown in Figure 14.12. We can note the changes in the signal flow, adder nodes, and branching nodes in the transposed IIR filters. By setting $ai = 0$ for $i = 1, 2, \dots, M-1$, we obtain the transposed structure of the FIR filter.

Both direct-form and transposed-form IIR filters are very sensitive to quantization errors because quantization errors in the feedback section are fed back and accumulated in the filter, which becomes severe when the filter order is high. A more robust structure for reducing quantization errors is to break down the high order IIR filter into a cascade of second-order IIR filters (or bi-quads). The general expression of cascading K bi-quads to form a $2X$-order IIR filter is

$$H(Z) = G \prod_{k=1}^{k} H_k(z) = G \prod_{k-1}^{k} \frac{b_{k0} + b_{k1}z^{-1} + b_{k2}z^{-2}}{1 + a_{k1}z^{-1} + a_{k2}z^{-2}} \tag{14.20}$$

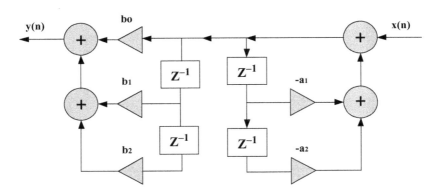

FIGURE 14.12 Transposed-direct-form-I.

where G is a gain. An example of cascading two bi-quads to form a fourth-order IIR filter is shown in Figure 14.12. Cascading a single first-order IIR filter with a series of filter transpositions may also be called *flow graph reversal*, and transposing a single-input/single-output (SISO) filter does not alter its transfer function. This fact can be derived as a consequence of *Mason's gain formula* for signal-flow graphs or *Tellegen's theorem* (which implies that a LTI signal-flow graph is *entered reciprocally* with its transpose). Transposed-direct-form-I implementation is a second order IIR digital filter. Note that the input signal comes in from the right, and the output is on the left.

14.6 TRANSPOSED-DIRECT-FORM-II

Figure 14.13 shows the *transposed-direct-form-II* structure. To facilitate comparison of the transposed with the original, the input and output signals remain "switched," so that signals generally flow right-to-left instead of the usual left-to-right.

Example 14.2

Find $H(z)$ for the difference equation:

$$y[n] = x[n] + x[n-1]$$

SOLUTION

The impulse response is: $\{h[n]\} = \{..., 0, \underline{1}, 1, 0,...\}$

$$\therefore H(z) = \sum_{n=-\infty}^{\infty} h[n]z^{-n} = \sum_{n=-0}^{1} h[n]z^{-n} = 1 + z^{-1}$$

Example 14.3

Find $H(z)$ for the recursive difference equation:

$$y[n] = a_0 x[n] + a_1 x[n-1] - b_1 y[n-1]$$

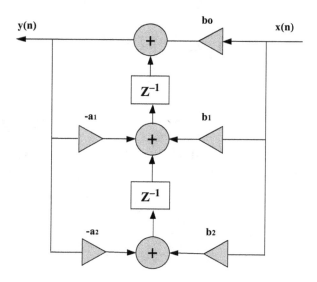

FIGURE 14.13   Transposed-direct-form-II.

SOLUTION

Remember that if $x[n] = z^n$ then $y[n] = H(z) z^n$, $y[n-1] = H(z) z^{n-1}$ etc. Substitute this into the difference equation to obtain:

$$H(z)z^n = a_0 z^n + a_1 z^{n-1} - b_1 H(z)z^{n-1}$$

Therefore,

$$H(z) = \frac{a_0 + a_1 z^{-1}}{1 + b_1 z^{-1}}$$

except when $z = -b_1$. When $z = -b_1$, $H(z) = \infty$.

By the same method, $H(z)$ for a general digital filter whose difference equation was given earlier is:

$$H(z) = \frac{a_0 + a_1 z^{-1} + a_2 z^{-2} + \cdots + a_N z^{-N}}{b_0 + b_1 z^{-1} + b_2 z^{-2} + \cdots + b_M z^{-M}} \text{ (with } b_0 = 1) \qquad (14.21)$$

Given $H(z)$ in this form, we can easily go back to its difference equation and hence its signal-flow graph, as illustrated by the following example.

Example 14.4

Give a signal-flow graph for the second-order digital filter with:

$$H(z) = \frac{a_0 + a_1 z^{-1} + a_2 z^{-2}}{1 + b_1 z^{-1} + b_2 z^{-2}}$$

SOLUTION

The difference equation is:

$$y[n] = a_0 x[n] + a_1 x[n-1] + a_2 x[n-2] - b_1 y[n-1] - b_2 y[n-2]$$

The signal-flow graph in Figure 14.14 is readily deduced from this difference equation. It is referred to as a second-order or "bi-quadratic" IIR section in "direct-form I."

Alternative signal-flow graphs can generally be found for a given difference equation. Considering again the "direct-form I" bi-quadratic section in Figure 14.14, re-ordering the two halves in this signal-flow graph provides Figure 14.15 which will have the same impulse response as the signal-flow graph in Figure 14.14. Now observe that may be simplified to the signal-flow diagram in

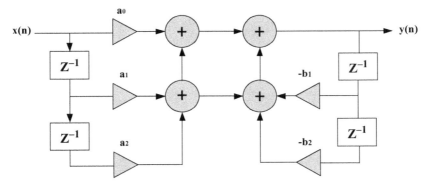

FIGURE 14.14　"Direct-form I" Bi-quadratic section.

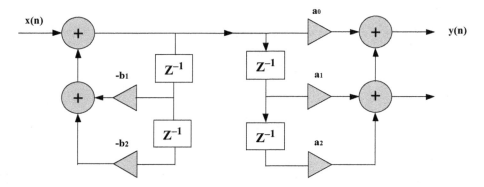

FIGURE 14.15   "Direct-form I" rearranged.

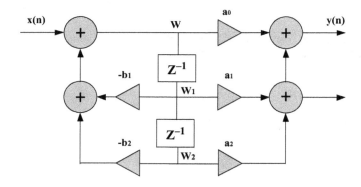

FIGURE 14.16   "Direct-Form II" Bi-quadratic section.

Figure 14.16 which is known as a "direct-form II" implementation of a bi-quadratic section. It has the minimum possible number of delay boxes and is said to be "canonical." Its system function is identical to that of the "direct-form I" signal-flow graph, and therefore it can be implemented any second-order bi-quadratic system function.

14.7   IMPLEMENTATION OF A NOTCH FILTER BY MATLAB

Assume, we want to design a fourth-order "notch" digital filter to eliminate an unwanted sinusoid at 800 Hz without severely affecting the rest of the signal. The sampling rate is FS = 10 kHz.

One simple way is to use the MATLAB function "butter" as follows:

```
FS=10000;
FL = 775; FU = 825;
[a b] = butter(2, [FL FU]/(FS/2),'stop');
a = [0.98 -3.43 4.96 -3.43 0.98];
b = [1 -3.47 4.96 -3.39 0.96];
freqz(a, b);
freqz(a, b, 512, FS);
axis([0 FS/2 -50 5]);
```

The frequency responses (gain and phase) produced by the final two MATLAB statements are as in Figure 14.17.

Since the Butterworth band-stop filter will have –3 dB gain at the two cut-off frequencies FL = 775 and FU = 825, the notch has "–3 dB frequency bandwidth": 25 + 25 = 50 Hz.

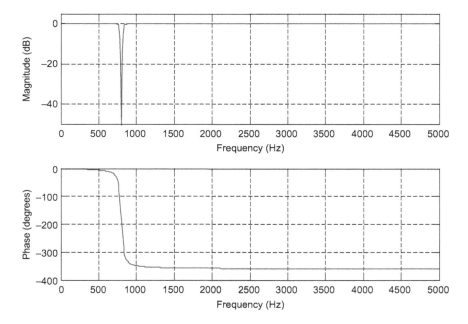

FIGURE 14.17   The frequency responses (gain and phase).

Now consider how to implement the fourth-order digital filter. The MATLAB function gave us:

```
a = [0.98 -3.43 4.96 -3.43 0.98]
b = [1 -3.47 4.96 -3.39 0.96]
```

The transfer (system) function is, therefore,

$$H(z) = \left(\frac{0.98 - 3.43z^{-1} + 4.96z^{-2} - 3.43z^{-3} + 0.98z^{-4}}{1 - 3.47z^{-1} + 4.96z^{-2} - 3.39z^{-3} + 0.96z^{-4}} \right)$$

A "direct-form II" implementation of the fourth-order notch filter would have the signal-flow graph shown if Figure 14.18.

This implementation works adequately in MATLAB. But "direct-form" IIR implementations of order greater than two are rarely used. Sensitivity to round-off error in coefficient values will be high. Also, the range of "intermediate" signals in the z^{-1} boxes will be high.

High word-length floating-point arithmetic hides this problem, but in fixed-point arithmetic, great difficulty occurs. Instead we use "cascaded bi-quad sections."

Given a fourth-order transfer function $H(z)$. Instead of the direct-form realization in Figure 14.19: we prefer to arrange two bi-quad sections, with a single leading multiplier G, as in Figure 14.20:

MATLAB will convert the fourth-order transfer function $H(z)$ to this new form. Do it as follows after getting a and b for the fourth-order transfer function, $H(z)$, as before:

```
[a b] = butter(2, [FL FU]/(FS/2),'stop');
[SOS G] = tf2sos(a,b)
```

- MATLAB responds with:

```
SOS = 1 -1.753 1 1 -1.722 0.9776
1 -1.753 1 1 -1.744 0.9785
G = 0.978
```

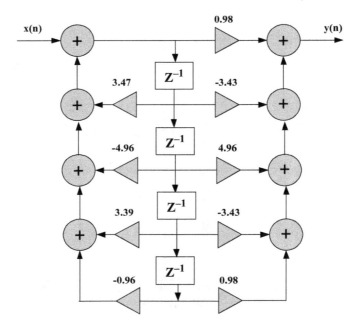

FIGURE 14.18   "Direct-Form II" implementation of the fourth-order notch filter.

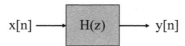

FIGURE 14.19   Equivalent fourth-order transfer function $H(z)$.

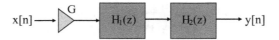

FIGURE 14.20   Arrange two bi-quad sections.

In MATLAB, "SOS" stands for "second-order section" (i.e., bi-quad) and the function "tf2SOS" converts the coefficients in arrays "a" and "b" to the new set of coefficients stored in array "SOS" and the constant G. The array SOS has two rows: one row for the first bi-quad section another row for the second bi-quad section. In each row, the first three terms specify the non-recursive part and the latter three terms define the recursive part. Therefore $H(z)$ may now be realized as shown in Figure 14.21.

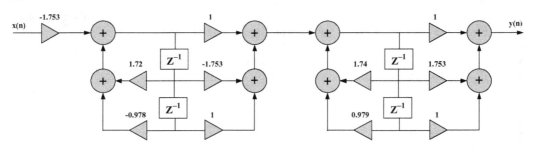

FIGURE 14.21   Fourth-order IIR notch filter recognized as two bi-quad (SOS) sections.

14.8 IMPLEMENTATION OF INFINITE-IMPULSE RESPONSE FILTERS

Since the analog-filter design is a well-developed technique, a systematic approach in designing a digital IIR filter is to design an analog IIR filter and then convert it into the equivalent digital filter. An alternate method is to create a digital IIR filter directly in the digital domain by using the algorithmic design procedure which solves a set of linear or nonlinear equations. The former approach is used to design typical low-pass, high-pass, band-pass, and band-stop filters. The latter approach is used to design filters with arbitrary frequency responses or constraints which have no analog prototype filters. We briefly explain the principles of these two design techniques in this section.

14.8.1 ANALOG-TO-DIGITAL FILTER DESIGN

An analog low-pass filter is first designed based on commonly used analog filters such as Butterworth, Chebyshev I and II, and elliptic filters. It is followed by either (1) performing frequency transformations in the analog domain and then applying analog-to-digital (or the s-plane to the z-plane) mapping or (2) applying analog-to-digital mapping before performing the frequency transformation in the digital domain. The frequency transformation is the process of converting a low-pass filter into another low-pass, band-pass, high-pass, or band-stop filter with different cut-off frequencies using some mapping functions. The frequency transformation can be carried out in either an analog or a digital domain.

14.8.2 BILINEAR TRANSFORMATION

A better and more commonly used the s-plane to the z-plane mapping is the bilinear transformation defined as

$$S = \frac{2(1-z^{-1})}{T(1+z^{-1})} \tag{14.22}$$

The bilinear transformation is a rational function that maps the left-half of the s-plane into the internal of the unit circle in the z-plane. It is a one-to-one mapping that maps the entire frequency along the $j\Omega$-axis from $-\infty$ to ∞ onto the unit circle ($-\pi$ to π) only once. This result of a nonlinear relationship between analog frequency and digital frequency is expressed as

$$\omega = 2\tan^{-1}\left(\frac{\Omega T}{2}\right) \tag{14.23}$$

LPF coefficients can be found, and the response can be plotted, as shown in Figure 14.22.

```
clc;
clear all;
close all;
rp=0.2; % passband ripple
rs=50; %stopband ripple
wp=1700; % passband freq
ws=3400; % stopband freq
fs=8000; %sampling freq
w1=2*wp/fs; w2=2*ws/fs;
[n,wn]=buttord(w1,w2,rp,rs,'s');
[b,a]=butter(n,wn,'low','s');
w=0:.1:pi;
[h,om]=freqs(b,a,w);
```

```
m=20*log10(abs(h));
an=angle(h);
figure
plot(om/pi,m);
title('IIR - LPF response ');
xlabel(' Normalized freq.');
ylabel('Gain (dB)');
```

Figure 14.22 shows the response of LPF using the `butter` command.
HPF coefficients can be found, and the response can be plotted, as shown in Figure 14.23.

```
clc;
clear all;
close all;
rp=0.2; %passband ripple
rs=100; %stopband ripple
wp=1500; % passband freq
ws=3000; % stopband freq
fs=10000;%sampling freq
w1=2*wp/fs;w2=2*ws/fs;
[n,wn]=buttord(w1,w2,rp,rs,'s');
[b,a]=butter(n,wn,'high','s');
w=0:.01:pi;
[h,om]=freqs(b,a,w);
m=20*log10(abs(h));
an=angle(h);
figure
plot(om/pi,m);
title('IIR-HPF response');
xlabel('Normalized freq.');
ylabel('Gain (dB)');
```

Figure 14.23 shows the response of HPF using the `butter` command.

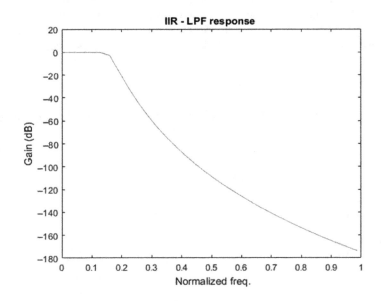

FIGURE 14.22   LPF using `butter` command.

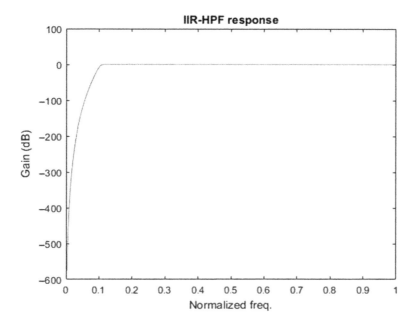

FIGURE 14.23   HPF using `butter` command.

PROBLEMS

14.1   Given a second-order transfer function

$$H(z) = \frac{5\left(1 - z^{-1}\right)}{\left(1 + z^{-1} + 3.36z^{-2}\right)}$$

Perform the filter realizations and write the difference equations using the following realizations.
(1) Direct-form I and direct-form II
(2) Cascade form via the first-order sections
(3) Parallel form via the first-order sections

14.2   Given a second-order transfer function

$$H(z) = \frac{0.5\left(1 - z^{-1}\right)}{\left(1 + 3z^{-1} + 2.5z^{-2}\right)}$$

Perform the filter realizations and write the difference equations using the following realizations.
(1) Direct-form I and direct-form II
(2) Cascade form via the first-order sections
(3) Parallel form via the first-order sections

14.3   Find $H(z)$ for the difference equation:

$$y[n] = x[n] + x[n-1]$$

14.4   Find $H(z)$ for the difference equation:

$$y[n] = x[n] + x[n-1]$$

14.5   Find $H(z)$ for the recursive difference equation:

$$y[n] = x[n] + x[n-1] + b_1\, y[n-1]$$

14.6   Find $H(z)$ for the recursive difference equation:

$$y[n] = 2a_0\, x[n] - a_1\, x[n-1] + b_1\, y[n-1]$$

14.7   Give a signal-flow graph for the second-order digital filter with:

$$H(z) = \frac{2 + 3z^{-1} + 4z^{-2}}{5 + 6z^{-1} + 7z^{-2}}$$

15 Implementation of FIR

In digital signal processing, a finite impulse response (FIR) filter is a filter whose impulse response or response to any finite length input is finite duration because it settles to zero in finite time. FIR filters contain as many poles as they have zeros. but all of the poles are located at the origin because all of the poles are located inside the unit circle, the FIR filter is ostensibly stable. This chapter will focus on the Implementation and design of FIR filter.

15.1 FINITE IMPULSE RESPONSE FILTER REPRESENTATION

A FIR filter is entirely specified by the following difference equation (input–output relationship):

$$y(n) = b_0 x(n) + b_1 x(n-1) + \cdots + b_M x(n-M) = \sum_{i=}^{M} b_i x(n-i) \tag{15.1}$$

The frequency response of an ideal low-pass filter can be shown in Figure 15.1a.
  Mathematically it is given by

$$H\left(e^{j\Omega}\right) = \begin{cases} 1, & 0 \le |\Omega| \le \Omega_c \\ 0, & \Omega_c \le |\Omega| \le \pi \end{cases} \tag{15.2}$$

The frequency Ω_c is the low-pass cut-off frequency. It can be shown that the corresponding impulse response of the ideal low-pass filter is, as shown in Figure 15.1b. A truncated part of the impulse response is shown (the actual response extends to infinity on both sides) from $n = -M$ to $n = M$. It is given by

$$h(n) = \begin{cases} \dfrac{\Omega_c}{\pi}, & n = 0 \\ \dfrac{\sin(\Omega_c n)}{\pi n} & n \ne 0 \end{cases} \tag{15.3}$$

The impulse response is symmetric to $n = 0$. In this case, the z-transform of the impulse response is given by

$$\begin{aligned} H(z) = {} & h(M)z^M + h(M-1)z^{M-1} + \cdots + h(1)z^1 \\ & + h(0) + h(1)z^{-1} + \cdots h(M-1)z^{-M+1} + h(M)z^{-M} \end{aligned} \tag{15.4}$$

To obtain a causal impulse response, we shift the non-causal impulse response by M samples, to yield the transfer function of a causal ideal low-pass FIR filter:

$$H(z) = b_0 + b_1 z^{-1} + b_2 z^{-2} + \cdots + b_{2M} z^{-2M} \tag{15.5}$$

where

$$b_n = h(n-M) \quad \text{for } n = 0,1,2,\ldots,2M. \tag{15.6}$$

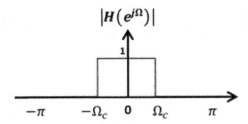

(a) Ideal Lowpass Frequency Response

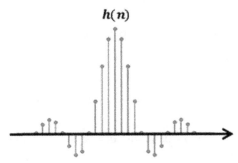

(b) Impulse Response of an Ideal Lowpass Filter

FIGURE 15.1 Finite Impulse Response. (a) Ideal low-pass frequency response. (b) impulse response of an ideal low-pass filter.

Also, we can obtain the design equations for the other types of FIR filters, such as high-pass, band-pass, and band-stop. Table 15.1 gives the formulas for these types of FIR filters for their filter coefficient calculations.

Example 15.1 Using an FIR low-pass filter

(a) Calculate the filter coefficients for a 3-tap with a cut-off frequency of 1,200 Hz and a sampling rate of 12,000 Hz using the Fourier transform method.
(b) Determine the transfer function and difference equation FIR system.
(c) Compute and plot the frequency response.

SOLUTION

Part (a): We first determine the normalized cut-off frequency

$$\Omega_c = 2\pi f_c T_s = 2\pi \times \frac{1200}{12000} = 0.2\,\pi \text{ radians}$$

In this case $2M + 1 = 3$, therefore, using Table 7.1,

$$h(n) = \begin{cases} \dfrac{\Omega_c}{\pi}, & n = 0 \\ \dfrac{\sin(\Omega_c n)}{\pi n} & -M \le n \le M \end{cases}$$

Therefore,

$$h(0) = \frac{\Omega_c}{\pi} = \frac{0.2\pi}{\pi} = 0.2$$

TABLE 15.1

Summary of Ideal Impulse Responses for Standard FIR Filters

Filter type	Ideal Impulse Response $h(n)$ (non-causal FIR coefficients)	

Low-pass

$$h(n) = \begin{cases} \dfrac{\Omega_c}{\pi}, & n = 0 \\ \dfrac{\sin(\Omega_c n)}{\pi n} & -M \le n \le M \end{cases}$$

High-pass

$$h(n) = \begin{cases} \dfrac{\pi - \Omega_c}{\pi}, & n = 0 \\ -\dfrac{\sin(\Omega_c n)}{\pi n} & -M \le n \le M \end{cases}$$

Band-pass

$$h(n) = \begin{cases} \dfrac{\Omega_H - \Omega_L}{\pi}, & n = 0 \\ \dfrac{\sin(\Omega_H n)}{\pi n} - \dfrac{\sin(\Omega_L n)}{\pi n} & -M \le n \le M \end{cases}$$

Band-stop

$$h(n) = \begin{cases} \dfrac{\pi - \Omega_H + \Omega_L}{\pi}, & n = 0 \\ -\dfrac{\sin(\Omega_H n)}{\pi n} + \dfrac{\sin(\Omega_L n)}{\pi n} & -M \le n \le M \end{cases}$$

Casual FIR-filter coefficients: shifting $h(n)$ to the right by samples.

Transfer function: $H(z) = b_0 + b\,z^{-1} + b_2 z^{-2} + \cdots + b_{2M} z^{-2M}$

Where $b_n = h(n - M)$ for $n = 0, 1, 2, ..., 2$

$$h(1) = \frac{\sin(\Omega_c)}{\pi} = \frac{\sin(0.2\pi)}{\pi} = 0.1871$$

using symmetry, $h(-1) = h(1) = 0.1871$

Delaying $h(n)$ by $M = 1$ samples, we get Filter Coefficients

$$b_0 = h(0 - 1) = h(-1) = 0.1871$$

$$b_1 = h(1 - 1) = h(0) = 0.2$$

$$b_2 = h(1 - 1) = h(1) = 0.1871$$

Part (b): Therefore, the transfer function in this case is

$$H(z) = b_0 + b_1 z^{-1} + b_2 z^{-2} = 0.1871 + 0.2 z^{-1} + 0.1871 z^{-2}$$

The difference equation is

$$y(n) = 0.1871 x(n) + 0.2 x(n - 1) + 0.1871 x(n - 2)$$

Part (c): The frequency response of the filter is

$$H\left(e^{j\Omega}\right) = 0.1871 + 0.2 e^{-j\Omega} + 0.1871 e^{-2j\Omega}$$

It can be written as

$$H\left(e^{j\Omega}\right) = e^{-j\Omega}\left(0.1871e^{j\Omega} + 0.2 + 0.1871e^{-2j\Omega}\right)$$

$$= e^{-j\Omega}\left(0.2 + 0.1871\left(e^{j\Omega} + e^{-2j\Omega}\right)\right)$$

$$= e^{-j\Omega}\left(0.2 + 0.1871 \times 2\cos\left(\Omega\right)\right)$$

$$= e^{-j\Omega}\left(0.2 + 0.3742\cos\left(\Omega\right)\right)$$

Thus, the magnitude frequency response is

$$\left|H\left(e^{j\Omega}\right)\right| = 0.2 + 0.3742\cos\left(\Omega\right)$$

And the phase response is

$$\angle H\left(e^{j\Omega}\right) = \begin{cases} -\Omega, & 0.2 + 0.3742\cos\left(\Omega\right) > 0 \\ -\Omega + \pi & 0.2 + 0.3742\cos\left(\Omega\right) < 0 \end{cases}$$

In FIR filters with symmetrical coefficients, the phase response has a linear behavior in the pass-band.

- It means that all frequency components of the filter input within the pass-band are subject to the same amount of time delay at the filter output. It is a requirement of applications in audio and speech filtering, where phase distortions need to be avoided.
- If the design method can't produce the symmetrical coefficients, then the resultant FIR filter does not have a linear-phase property which leads to distortions in the filtered signal. This distortion is due to the nonlinear phase shown with the example below.

$$x(n) = x_1(n) + x_2(n) = \sin(0.05\pi n)u(n) + \sin(0.15\pi n)u(n)$$

$$y_1(n) = \sin\left(0.05\pi(n-8)\right) + \sin\left(0.15\pi(n-8)\right) \text{ Linear Phase}$$

$$y_2(n) = \sin\left(0.05\pi(n-\pi/2)\right) + \sin\left(0.15\pi(n-\pi/2)\right) \text{ Non-Linear Phase}$$

The FIR filter has a good phase property, but it does not give an acceptable magnitude frequency response. The result with a 17-tap low-pass filter is also shown below.

- There are oscillations (ripple) in the pass-band (main lobe) and stop-band (side lobe) of the magnitude frequency response. This is due to Gibbs oscillations, originating from the abrupt truncation of the infinite-impulse-response of the low-pass filter.
- This behavior can be avoided with the help of windowing.

Example 15.2

(a) Calculate the filter coefficients for a 5-tap FIR band-pass filter with a lower cut-off frequency of 2,000 Hz and an upper cut-off frequency of 2,400 at a sampling rate of 8,000 Hz.
(b) Determine the transfer function and plot the frequency responses with MATLAB.

SOLUTION

Part (a): We first determine the normalized cut-off frequencies

$$\Omega_L = \frac{2\pi f_L}{f_s} = 2\pi \times \frac{2000}{8000} = 0.5\pi \text{ radians}$$

$$\Omega_H = \frac{2\pi f_L}{f_s} = 2\pi \times \frac{2400}{8000} = 0.6\pi \text{ radians}$$

In this case $2M + 1 = 5$, therefore, from Table 7.1,

$$h(n) = \begin{cases} \dfrac{\Omega_H - \Omega_L}{\pi}, & n = 0 \\ \dfrac{\sin(\Omega_H n)}{n\pi} - \dfrac{\sin(\Omega_L n)}{n\pi}, & -2 \le n \le 2 \end{cases}$$

The noncausal FIR coefficients are

$$h(0) = \frac{\Omega_H - \Omega_L}{\pi} = \frac{0.6\pi - 0.5\pi}{\pi} = 0.1$$

$$h(1) = \frac{\sin(\Omega_H \times 1)}{\pi \times 1} - \frac{\sin(\Omega_L \times 1)}{\pi \times 1} = \frac{\sin(0.6\pi)}{\pi} - \frac{\sin(0.5\pi)}{\pi} = -0.01558$$

$$h(2) = \frac{\sin(\Omega_H \times 2)}{\pi \times 2} - \frac{\sin(\Omega_L \times 2)}{\pi \times 2} = \frac{\sin(1.2\pi)}{2\pi} - \frac{\sin(1.0\pi)}{2\pi} = -0.09355$$

Using the symmetry property

$$h(-1) = h(1) = -0.0158$$

$$h(-2) = h(2) = -0.09355$$

Thus, the filter coefficients are obtained by delaying by $M = 2$ samples, as Filter Coefficients

$$b_0 = h(0 - 2) = h(-2) = -0.09355$$

$$b_1 = h(1 - 2) = h(-1) = -0.01558$$

$$b_2 = h(2 - 2) = h(0) = 0.1$$

$$b_3 = h(3 - 2) = h(1) = -0.01558$$

$$b_4 = h(4 - 2) = h(2) = -0.09355$$

Part (b): Therefore, the transfer function in this case is

$$H(z) = b_0 + b_1 z^{-1} + b_2 z^{-2} + b_3 z^{-3} + b_4 z^{-4}$$

$$= -0.09355 - 0.01558 z^{-1} + 0.1 z^{-2}$$

$$- 0.01558 z^{-3} - 0.09355 z^{-4}$$

The difference equation is

$$y(n) = -0.09355\, x(n) - 0.01558\, x(n-1) + 0.1 x(n-2)$$

$$- 0.01558\, x(n-3) - 0.09355\, x(n-4)$$

Part (c): The frequency response of the filter is

$$H(e^{j\Omega}) = -0.09355 - 0.01558 e^{-j\Omega} + 0.1 e^{-2j\Omega}$$

$$- 0.01558 e^{-3j\Omega} - 0.09355 e^{-4j\Omega}$$

It can be written as

$$H\left(e^{j\Omega}\right) = e^{-2j\Omega}\left(-0.09355\,e^{2j\Omega} - 0.01558\,e^{j\Omega} + 0.1 - 0.01558\,e^{-j\Omega} - 0.09355\,e^{-2j\Omega}\right)$$

$$= e^{-2j\Omega}\left(0.1 - 0.01558\left(e^{j\Omega} + e^{-j\Omega}\right) - 0.09355\left(e^{2j\Omega} + e^{-2j\Omega}\right)\right)$$

$$= e^{-2j\Omega}\left(0.1 - 0.01558\left(2\cos\left(\Omega\right)\right) - 0.09355\left(2\cos\left(2\Omega\right)\right)\right)$$

$$= e^{-2j\Omega}\left(0.1 - 0.03116\cos\left(\Omega\right) - 0.1871\cos\left(2\Omega\right)\right)$$

15.2   WINDOW METHOD

We have seen Gibbs oscillations are produced in the pass-band and stop-band in the design of FIR filters which are not desirable features of the FIR filter. To solve this problem, the window method is developed.

The Gibbs oscillations mainly result due to the abrupt truncation of the infinite-length coefficient sequence of the impulse response of the FIR filter. A window function is a symmetrical function which can gradually weigh the designed FIR coefficients down to zeros at both ends of for the range $-M \leq n \leq M$. Applying the window sequence to the filter coefficients gives

$$h_w(n) = h(n)\cdot w(n) \tag{15.7}$$

where, $w(n)$ designates the window function. The commonly used window functions in FIR filters are:

---

1. **Rectangular Window:**

$$w_{\text{rec}}(n) = 1, \quad -M \leq n \leq M \tag{15.8}$$

2. **Triangular (Bartlett) Window:**

$$w_{\text{tri}}(n) = 1 - \frac{|n|}{M}, \quad -M \leq n \leq M \tag{15.9}$$

3. **Hanning Window:**

$$w_{\text{han}}(n) = 0.5 + 0.5\cos\left(\frac{n\pi}{M}\right), \quad -M \leq n \leq M \tag{15.10}$$

4. **Hamming Window:**

$$w_{\text{ham}}(n) = 0.54 + 0.46\cos\left(\frac{n\pi}{M}\right), \quad -M \leq n \leq M \tag{15.11}$$

5. **Blackman Window:**

$$w_{\text{black}}(n) = 0.42 + 0.5\cos\left(\frac{n\pi}{M}\right) + 0.08\cos\left(\frac{2n\pi}{M}\right), \quad -M \leq n \leq M \tag{15.12}$$

Example 15.3

(a) Calculate the filter coefficients for a 3-tap FIR low-pass filter with a cut-off frequency of 1,000 Hz and a sampling rate of 10,000 Hz using the Hamming window function.
(b) Determine the transfer function and difference equation of the designed FIR system.
(c) Compute and plot the frequency response.

SOLUTION

Part (a): We first determine the normalized cut-off frequency

$$\Omega_c = 2\pi f_c T_s = 2\pi \times \frac{1000}{10000} = 0.2\pi \, \text{radians}$$

In this case $2M+1=3$, therefore, using Table 7.1,

$$h(n) = \begin{cases} \dfrac{\Omega_c}{\pi}, & n = 0 \\ \dfrac{\sin(\Omega_c n)}{\pi n} & -M \le n \le M \end{cases}$$

Therefore,

$$h(0) = \frac{\Omega_c}{\pi} = \frac{0.2\pi}{\pi} = 0.2$$

$$h(1) = \frac{\sin(\Omega_c)}{\pi} = \frac{\sin(0.2\pi)}{\pi} = 0.1871$$

$$h(-1) = h(1) = 0.1871 \, \text{(using symmetry)}$$

The windowed impulse response is calculated as

$$h_w(0) = h(0)w_{\text{ham}}(0) = 0.2 \times 1.0 = 0.2$$

$$h_w(1) = h(1)w_{\text{ham}}(1) = 0.1871 \times 0.08 = 0.01497$$

$$h_w(-1) = h(-1)w_{\text{ham}}(-1) = 0.1871 \times 0.08 = 0.01497$$

Delaying $h(n)$ by $M = 1$ samples, we get $b_0 = h_w(-1) = 0.01497$, $b_1 = h_w(0) = 0.2$, and $b_2 = h_w(1) = 0.01497$.

Part (b): Therefore, the transfer function in this case is

$$H(z) = b_0 + b_1 z^{-1} + b_2 z^{-2} = 0.01497 + 0.2z^{-1} + 0.01497z^{-2}$$

The difference equation is

$$y(n) = 0.01497x(n) + 0.2x(n-1) + 0.01497x(n-2)$$

Part (c): The frequency response of the filter is

$$H\left(e^{j\Omega}\right) = 0.01497 + 0.2e^{-j\Omega} + 0.01497e^{-2j\Omega}$$

It can be written as

$$H\left(e^{j\Omega}\right) = e^{-j\Omega}\left(0.01497\,e^{j\Omega} + 0.2 + 0.01497\,e^{-2j\Omega}\right)$$

$$= e^{-j\Omega}\left(0.2 + 0.01497\left(e^{j\Omega} + e^{-2j\Omega}\right)\right)$$

$$= e^{-j\Omega}\left(0.2 + 0.001497 \times 2\cos(\Omega)\right)$$

$$= e^{-j\Omega}\left(0.2 + 0.02994\cos(\Omega)\right)$$

Thus, the magnitude frequency response is

$$\left|H\left(e^{j\Omega}\right)\right| = 0.2 + 0.02994\cos(\Omega)$$

And the phase response is

$$\angle H\left(e^{j\Omega}\right) = \begin{cases} -\Omega, & 0.2 + 0.02994\cos(\Omega) > 0 \\ -\Omega + \pi & 0.2 + 0.02994\cos(\Omega) < 0 \end{cases}$$

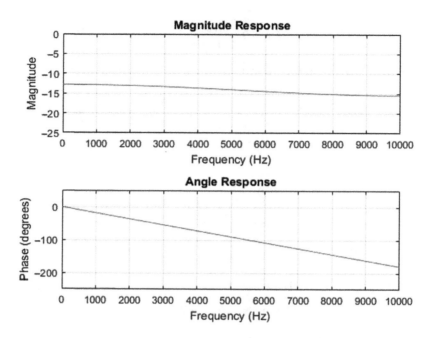

FIGURE 15.2 Frequency response in Example 15.3.

A MATLAB program to plot the frequency response is given in the following (Figure 15.2).

```
clc;
clear all;
fs = 10000;
B = [0.01497 0.2 0.01497];
A = [1];
[h w] = freqz(B, A, 1024);
f1 = w*fs/(pi);
Angle = 180*unwrap(angle(h))/pi;
figure
subplot(2,1,1), plot(f1, 20*log10(abs(h)));
grid on;
axis([0 fs -25 0]);
xlabel('Frequency (Hz)');
ylabel('Magnitude');
title('Magnitude Response')'
subplot(2,1,2),
plot(f1, Angle);
grid on;
axis([0 fs -250 50]);
xlabel('Frequency (Hz)');
ylabel('Phase (degrees)');
title('Angle Response')'
```

Example 15.4

To implement the FIR filter for a given sequence by using windowing techniques. A Finite Impulse Response (FIR) filter is a discrete linear time-invariant system whose output is based on the weighted summation of a finite number of past inputs.

A FIR transversal filter structure can be obtained directly from the equation for discrete-time convolution.

$$y\,[n] = \sum_{k=0}^{N-1} X(k)\,h(n-k)\ 0 < n < N-1$$

In this equation, $x(k)$ and $y(n)$ represent the input to and output from the filter at time n. $h(n-k)$ are the transversal filter coefficients at time n. These coefficients are generated by using FDS (filter-design software or digital filter-design package).

Types of windows available are Rectangular, Barlett, Hamming, Hanning, Blackman, etc. This FIR filter is an all-zero filter.

```
% window function rectangular

clc;
clear all;
close all;
rp=0.02; % passband ripple
rs=0.01; % stopband ripple
fp=1000; %passband freq
fs=2000; %stopband freq
f=10000; %sampling freq
wp=2*fp/f;
ws=2*fs/f;
num=-20*log10(sqrt(rp*rs))-13;
dem=14.6*(fs-fp)/f;
n=ceil(num/dem);
n1=n+1;
if(rem(n,2)~=0)
n1=n;
n=n-1;
end
y=rectwin(n1);
%LPF
b=fir1(n,wp,y);
[h,o]=freqz(b,1,256);
m=20*log10(abs(h));
subplot(2,1,1);plot(o/pi,m);
title('LPF');
ylabel('Gain in dB');
xlabel(' Normalized frequency');
%HPF
b=fir1(n,wp,'high',y);
[h,o]=freqz(b,1,256);
m=20*log10(abs(h));
subplot(2,1,2);
plot(o/pi,m);
title('HPF');
ylabel('Gain in dB');
xlabel(' Normalized frequency');
```

Figure 15.3 shows the response of a window function rectangular FIR filter for a low- and high-pass filter.

```
% window function triangular
clc;
clear all;
```

```
close all;
rp=0.02; % passband ripple
rs=0.01; % stopband ripple
fp=1000; %passband freq
fs=2000; %stopband freq
f=10000; %sampling freq
wp=2*fp/f;
ws=2*fs/f;
num=-20*log10(sqrt(rp*rs))-13;
dem=14.6*(fs-fp)/f;
n=ceil(num/dem);
n1=n+1;
if(rem(n,2)~=0)
n1=n;
n=n-1;
end
y=triang(n1);
%LPF
b=fir1(n,wp,y);
[h,o]=freqz(b,1,256);
m=20*log10(abs(h));
subplot(2,1,1);plot(o/pi,m);
title('LPF');
ylabel('Gain(dB)');
xlabel('Normalized frequency');
%HPF
b=fir1(n,wp,'high',y);
[h,o]=freqz(b,1,256);
m=20*log10(abs(h));
```

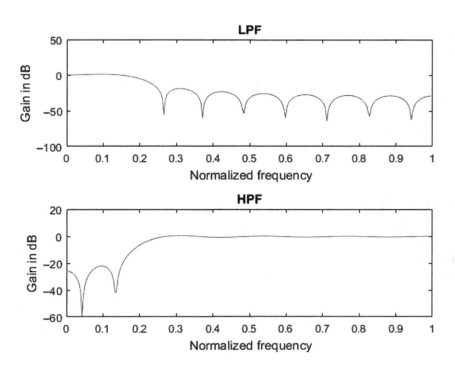

FIGURE 15.3 Response of window function rectangular FIR filter.

```
subplot(2,1,2);plot(o/pi,m);
title('HPF');
ylabel('Gain(dB)');
xlabel('Normalized frequency');
```

Figure 15.4 shows the response of a window function triangular FIR filter for a low- and high-pass filter.

Example 15.5

(a) Design a 5-tap FIR band-reject filter with a lower cut-off frequency of 2,000 Hz, an upper cut-off frequency of 2,400 Hz and a sampling rate of 8,000 Hz using the Hamming window method.
(b) Determine the transfer function.

SOLUTION

Part (a): We first determine the normalized cut-off frequencies

$$\Omega_L = 2\pi f_L T_s = 2\pi \times \frac{2000}{8000} = 0.5\pi \text{ radians}$$

$$\Omega_H = 2\pi f_H T_s = 2\pi \times \frac{2400}{8000} = 0.6\pi \text{ radians}$$

In this case $2M + 1 = 5$, therefore, using Table 7.1,

$$h(n) = \begin{cases} \dfrac{\pi - \Omega_H - \Omega_L}{\pi} & n = 0 \\[3mm] \left| \dfrac{\sin(\Omega_H n)}{n\pi} + \dfrac{\sin(\Omega_H n)}{n\pi} \right| & -2 \le n \le 2 \end{cases}$$

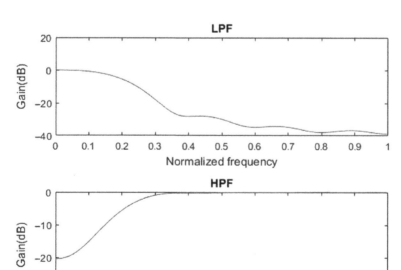

FIGURE 15.4 Response of window function triangular FIR filter.

Therefore,

$$h(0) = \frac{\pi - \Omega_H - \Omega_L}{\pi} = \frac{\pi - 0.6\pi - 0.5\pi}{\pi} = 0.9$$

$$h(1) = -\frac{\sin(\Omega_H)}{\pi} + \frac{\sin(\Omega_H)}{\pi} = -\frac{\sin(0.6\pi)}{\pi} + \frac{\sin(0.5\pi)}{\pi} = 0.01558$$

$$h(2) = -\frac{\sin(2\Omega_H)}{2\pi} + \frac{\sin(\Omega_H 2)}{2\pi} = -\frac{\sin(1.2\pi)}{\pi} + \frac{\sin(\pi)}{\pi} = 0.09355$$

$$h(-1) = h(1) = 0.01558 \text{ (using symmetry)}$$

$$h(-2) = h(2) = 0.09355 \text{ (using symmetry)}$$

Applying the Hamming window function, we have

$$w_{\text{ham}}(0) = 0.54 + 0.46 \cos\left(\frac{0 \times \pi}{2}\right) = 1.0$$

$$w_{\text{ham}}(1) = 0.54 + 0.46 \cos\left(\frac{1 \times \pi}{2}\right) = 0.54$$

$$w_{\text{ham}}(2) = 0.54 + 0.46 \cos\left(\frac{2 \times \pi}{2}\right) = 0.08$$

$$w_{\text{ham}}(-1) = w_{\text{ham}}(1) = 0.54 \text{ (using symmetry)}$$

$$w_{\text{ham}}(-2) = w_{\text{ham}}(2) = 0.08 \text{ (using symmetry)}$$

The windowed impulse response is calculated as

$$h_w(0) = h(0)w_{\text{ham}}(0) = 0.9 \times 1.0 = 0.9$$

$$h_w(1) = h(1)w_{\text{ham}}(1) = 0.01558 \times 0.54 = 0.00841$$

$$h_w(2) = h(2)w_{\text{ham}}(2) = 0.09355 \times 0.08 = 0.00748$$

$$h_w(-1) = h(-1)w_{\text{ham}}(-1) = 0.01558 \times 0.54 = 0.00841$$

$$h_w(-2) = h(-2)w_{\text{ham}}(-2) = 0.09355 \times 0.08 = 0.00748$$

Delaying $h(n)$ by $M = 2$ samples, we get $b_0 = h_w(-2) = 0.00748$, $b_1 = h_w(-1) = 0.00841$, $b_2 = h_w(0) = 0.9$, $b_3 = h_w(1) = 0.00841$ and $b_4 = h_w(2) = 0.00748$.

Part (b): Therefore, the transfer function in this case is

$$H(z) = b_0 + b_1 z^{-1} + b_2 z^{-2} + b_3 z^{-3} + b_4 z^{-4}$$

$$= 0.00748 + 0.00841 z^{-1} + 0.9 z^{-2}$$

$$+ 0.00841 z^{-3} + 0.00748 z^{-4}$$

15.3 FIR-FILTER LENGTH ESTIMATION USING WINDOW FUNCTIONS

Given the required pass-band ripples specification and stop-band attenuation, the appropriate window length can be estimated based on the performances of the window functions. For illustrative purpose, we use the low-pass filter frequency domain specification (the same can be extended to other types of filter specifications).

The normalized transition band frequency is defined as

$$\Delta f = \left| f_{\text{stop}} - f_{\text{pass}} \right| / f_s \tag{15.13}$$

Based on this, the FIR-filter lengths for various window functions are given in Table 15.2 below. It can be noted that the cut-off frequency is determined by

$$f_c = \left(f_{\text{stop}} - f_{\text{pass}} \right) / 2 \tag{15.14}$$

The pass-band ripple is defined as

$$\delta_p \, dB = 20 \cdot \log_{10} \left(1 + \delta_p \right) \tag{15.15}$$

while the stop-band attenuation is defined as (Table 15.2)

$$\delta_s \, dB = -20 \cdot \log_{10} \left(\delta_s \right) \tag{15.16}$$

Example 15.6

Design a band-pass FIR filter with the following specifications:

    Lower stop-band = 0 – 600 Hz
    Pass-band = 1,500 – 2,400 Hz
    Upper stop-band = 3,500 – 4,000 Hz
    Stop-band attenuation = 50 dB
    Pass-band ripple = 0.05dB
    Sampling rate = 10,000 Hz

Determine the FIR-filter length and the cut-off frequency to be used in the design equation.

SOLUTION

We first determine the normalized transition band

$$\Delta f_1 = \frac{\left| 1500 - 600 \right|}{10000} = 0.09$$

TABLE 15.2

Filter Length Estimation Using Window ($\Delta f = \left| f_{\text{stop}} - f_{\text{pass}} \right| / f_s$)

Window Type	Window Function $w(n),\ -M \leq n \leq M$	Window Length	Pass-band Ripple (dB)	Stop-band Attenuation (dB)
Rectangular	1	$N = 0.9/\Delta f$	0.7416	21
Hanning	$0.5 + 0.5\cos\left(\dfrac{n\pi}{M}\right)$	$N = 3.1 / \Delta f$	0.0546	44
Hamming	$0.54 + 0.46\cos\left(\dfrac{n\pi}{M}\right)$	$N = 3.3 / \Delta f$	0.0194	53
Blackman	$0.42 + 0.5\cos\left(\dfrac{n\pi}{M}\right) + 0.08\cos\left(\dfrac{2n\pi}{M}\right)$	$N = 5.5 / \Delta f$	0.0017	74

$$\Delta f_2 = \frac{|3500 - 2400|}{10000} = 0.11$$

The filter lengths based on above transition bands (for Hamming window) are

$$N_1 = \frac{3.3}{0.09} = 37$$

$$N_2 = \frac{3.3}{0.11} = 30$$

The nearest higher odd N is chosen as for the Hamming window

$$N = 37$$

The lower and higher cut-off frequencies for the band-pass filter will be

$$f_1 = \frac{1500 + 600}{2} = 1050\,\text{Hz}$$

$$f_1 = \frac{3500 + 2400}{2} = 2950\,\text{Hz}$$

The normalized lower and higher cut-off frequencies for the band-pass filter will be

$$\Omega_L = 2\pi f_L T_s = 2\pi \times \frac{1050}{10000} = 0.21\pi\,\text{radians}$$

$$\Omega_H = 2\pi f_H T_s = 2\pi \times \frac{2950}{10000} = 0.59\pi\,\text{radians}$$

In this case, $2M + 1 = 37$ will be the number of taps for the band-pass filter.

A MATLAB program to plot the frequency response is given in the following (Figure 15.5).

```
clc;
clear all;
Ntap = 37;
Ftype = 3;
WL = 0.21*pi;
WH = 0.59*pi;
Wtype = 4;
fs = 10000;
B=firwd(Ntap,Ftype,WL,WH,Wtype);
A = 1;
omega = 0:0.1:pi;
f = omega*fs/(2*pi);
[hz, w] = freqz(B, A, omega);
magnitude = 20*log10(abs(hz));
phase = 180*unwrap(angle(hz))/pi;
figure,
subplot(2,1,1);
plot(f, magnitude);
grid on;
xlabel('Frequency (Hz)');
ylabel('Magnitude response(dB)');
subplot(2,1,2); plot(f,phase);
grid on;
xlabel('Frequency (Hz)');
ylabel('Phase response(degrees)');
```

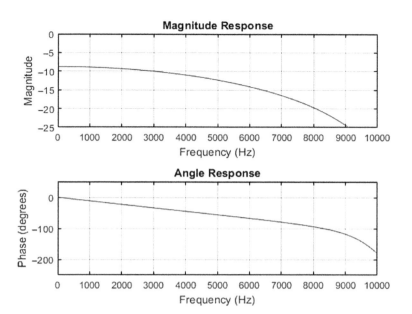

FIGURE 15.5 Frequency response in Example 15.6.

The FIR have a finite number of elements and in $h(n)$ they are

$$H(n) = \left[h(0), h(1), h(2), \ldots, h(N+1) \right]$$

$$H(z) = h(0) + h(1)z^{-1} + h(2)z^{-2} + \cdots + h(N+1)z^{-(N+1)} \tag{15.17}$$

$$H(z) = \sum_{k=0}^{N-1} h(k)z^{-k} = \frac{Y(z)}{X(z)} \tag{15.18}$$

$$Y(z) = X(z)\left[h(0) + h(1)z^{-1} + h(2)z^{-2} + \cdots + h(N-1)z^{-(N-1)} \right]$$

$$Y(n) = h(0)x(n) + h(1)x(n-1) + h(2)x(n-2) + \cdots \tag{15.19}$$

The system equation consists of only zeros, and has no feedback (Figure 15.6).
The digital filter above consists of three parts, which are

1. Delay represented by z^{-1}.
2. Multiplication.
3. Addition.

Example 15.7

Find the output samples sequence values from the digital filter shown in Figure 15.7 if the input sequence $x(n) = [1\ 1\ 1\ 1]$ (Figure 15.8).

SOLUTION

At $n = 0$. $x(0) = 1$
At $n = 1$. $x(1) = 1$

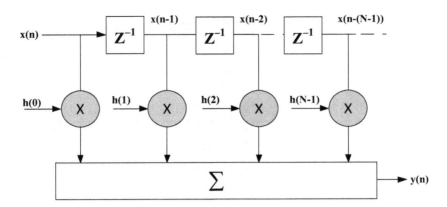

FIGURE 15.6   Structure of FIR filter.

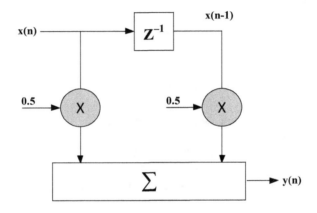

FIGURE 15.7   System for Example 15.7.

At $n = 2$. $x(2) = 1$
At $n = 3$. $x(3) = 1$
$y(n) = [0.5\ 1\ 1\ 1]$

Example 15.8

Find the FFT for the input and output of the digital filter as shown in Figure 15.9 for the input sequence of $x(n) = [1\ 1\ 1\ 1]$.

SOLUTION

1) FFT for $x(n)$ is given in Figures 15.10 and 15.11
2) FFT for $y(n)$, must find the output $y(n)$ from the digital filter as follows Figure 15.12.

   So, $y(n) = [0.5\ 0\ 0\ 0]$
   Now use the FFT for $y(n)$ to find $Y(k)$ Figure 15.13
   So, $Y(k) = [0.5\ 0.5\ 0.5\ 0.5]$ Figure 15.14

Example 15.9

Design a low-pass FIR filter with the following specifications:

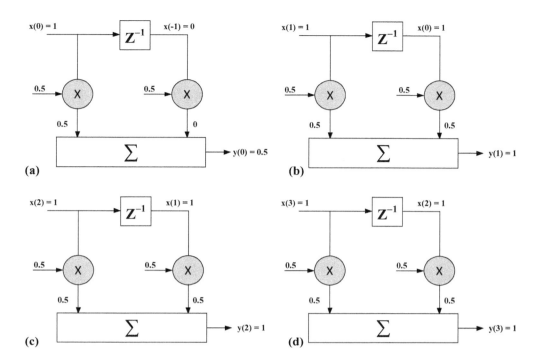

FIGURE 15.8   System for steps of the solution of Example 15.7.

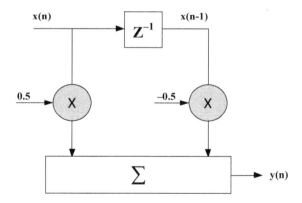

FIGURE 15.9   System of Example 15.8.

Cut-off frequency = 1 kHz
Stop-band attenuation = 28 dB at frequency 2 kHz
Sampling rate = 8 kHz use Hanning window

SOLUTION

The normalized lower and higher cut-off frequencies for the band-pass filter will be

$$\omega_{1new} = 2\pi f_L T_s = 2\pi \times \frac{1000}{8000} = 0.25\pi \text{ radians}$$

$$\omega_{2new} = 2\pi f_H T_s = 2\pi \times \frac{2000}{8000} = 0.75\pi \text{ radians}$$

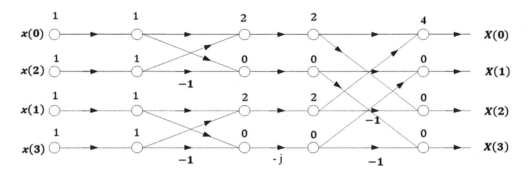

FIGURE 15.10   FFT for $x(n)$ of Example 15.8.

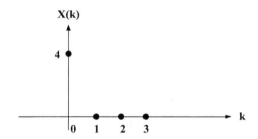

FIGURE 15.11   FFT output for $x(n)$ of Example 15.8.

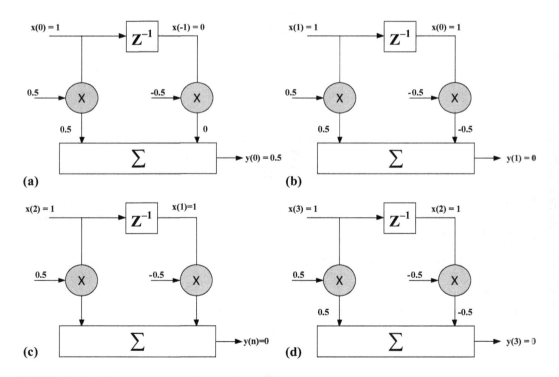

FIGURE 15.12   System for steps of the solution of Example 15.8.

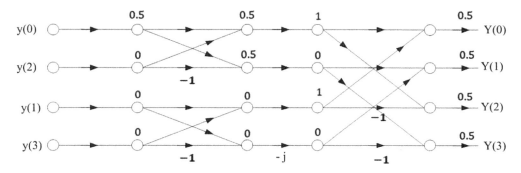

FIGURE 15.13 FFT for $y(n)$ of Example 15.8.

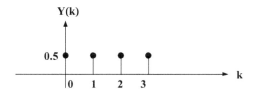

FIGURE 15.14 FFT output for $y(n)$ of Example 15.8.

$$\omega_{1new} = \omega_c$$

$$N = \frac{2\pi \times \pi}{0.5\pi - 0.25\pi} = 33$$

$$2M + 1 = 33$$

$$M = 16$$

$$w_{han}(n) = 0.5 + 0.5\cos\left(\frac{n\pi}{M}\right), \quad -M \le n \le M$$

For LPF

$$h(n) = \begin{cases} \dfrac{\Omega_c}{\pi}, & n = 0 \\ \dfrac{\sin(\Omega_c n)}{\pi n} & -M \le n \le M \end{cases}$$

PROBLEMS

15.1 Calculate the filter coefficients for a 3-tap FIR low-pass filter with a cut-off frequency of 1,000 Hz and a sampling rate of 10,000 Hz using the Fourier Transform method.
(a) Determine the transfer function and difference equation of the designed FIR system.
(b) Compute and plot the frequency response.

15.2 Calculate the filter coefficients for a 3-tap FIR low-pass filter with a cut-off frequency of 450 Hz and a sampling rate of 5,000 Hz using the Fourier Transform method.
(a) Determine the transfer function and difference equation of the designed FIR system.
(b) Compute and plot the frequency response.

15.3 (a) Calculate the filter coefficients for a 5-tap FIR band-pass filter with a lower cut-off frequency of 1,800 Hz and an upper cut-off frequency of 2,000 at a sampling rate of 10,000 Hz.
      (b) Determine the transfer function and plot the frequency responses with MATLAB.

15.4 (a) Calculate the filter coefficients for a 5-tap FIR band-pass filter with a lower cut-off frequency of 2,000 Hz and an upper cut-off frequency of 2,800 at a sampling rate of 5,000 Hz.
      (b) Determine the transfer function and plot the frequency responses with MATLAB.

15.5 (a) Calculate the filter coefficients for a 3-tap FIR low-pass filter with a cut-off frequency of 4,400 Hz and a sampling rate of 8,500 Hz using the Hamming window function.
      (b) Determine the transfer function and difference equation of the designed FIR system.
      (c) Compute and plot the frequency response.

15.6 (a) Design a 5-tap FIR band-reject filter with a lower cut-off frequency of 1,000 Hz, an upper cut-off frequency of 1,400 Hz and a sampling rate of 8,000 Hz using the Hamming window method.
      (b) Determine the transfer function.

15.7 Design a band-pass FIR filter with the following specifications:

Lower Stop-band = $0 - 550$ Hz
Pass-band = $1,650 - 2,350$ Hz
Upper stop-band = $3,550 - 4,100$ Hz
Stop-band attenuation = 52 dB
Pass-band ripple = 0.04 dB
Sampling rate = 10,000 Hz

Determine the FIR-filter length and the cut-off frequency to be used in the design equation.

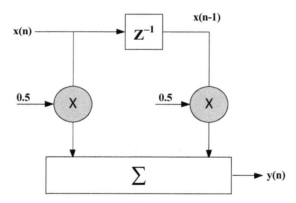

FIGURE 15.15   System for problem 15.8.

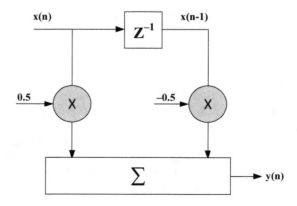

FIGURE 15.16   System for problem 15.9.

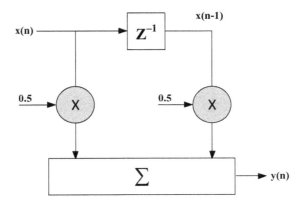

FIGURE 15.17 System for problem 15.10.

15.8   Find the output samples sequence values from the digital filter shown in Figure 15.15 if the input sequence $x(n) = [1\ 0\ 0\ 1]$.

15.9   Find the output samples sequence values from the digital filter shown in Figure 15.16 if the input sequence $x(n) = [0\ 1\ 1\ 0]$.

15.10 Find the FFT for the input and output of the digital filter as shown in Figure 15.17 for the input sequence of $x(n) = [1\ 0\ 1\ 0]$.

15.11 Design a low-pass FIR filter with the following specifications:

Cut-off frequency = 1.2 kHz
Stop-band attenuation = 27 dB at frequency 2 kHz
Sampling rate = 8.5 kHz use Hamming window

16 Digital Filter Design

16.1 IIR FILTER DESIGN

There are many methods for designing an IIR filter, but one of the simplest ways is to use an analog-filter design then with a suitable transformation the function $H(s)$ is transformed to digital in the z-domain $H(z)$.

16.1.1 ANALOG-FILTER DESIGN

These are classified into the Butterworth filter and the Chebyshev filter.

1) Butterworth filter

a) low-pass filter
To design a low-pass filter use the following steps:

Step 1: From the characteristics of the filter in the frequency domain as in Figure 16.1, the order of the filter is calculated from

$$n = \frac{\log_{10}\left[\dfrac{10^{-\frac{k_1}{10}}-1}{10^{-\frac{k_2}{10}}-1}\right]}{2\log_{10}\left[\dfrac{\omega_1}{\omega_2}\right]} \tag{16.1}$$

Step 2: From Table 16.1, find how the function $B(s)$ is related to the order n:
Step 3: Find the cut-off frequency

$$\omega_c = \frac{\omega_1}{\left(10^{-\frac{k_1}{10}}-1\right)^{1/2n}} \tag{16.2}$$

Step 4: Find the transfer function of the filter

$$H(s) = \frac{1}{B(s)}\Bigg|_{s=\frac{s}{\omega_c}} \tag{16.3}$$

Example 16.1

Design analog LPF that has pass-band attenuation of 2 dB at 15 rad/sec and stop-band attenuation of 10 dB at 30 rad/sec (Figure 16.2).

$$n = \frac{\log_{10}\left[\dfrac{10^{-\frac{k_1}{10}}-1}{10^{-\frac{k_2}{10}}-1}\right]}{2\log_{10}\left[\dfrac{\omega_1}{\omega_2}\right]} = 2$$

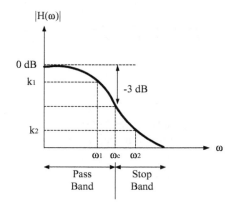

FIGURE 16.1 Low-pass Butterworth filter response.

TABLE 16.1
Binominal of Filter Denominator

n	$B(s)$
1	$S + 1$
2	$s^2 + \sqrt{2}s + 1$
3	$(s+1)(s^2 + s + 1)$
4	$(s^2 + 0.7653s + 1)(s^2 + 1.8477s + 1)$

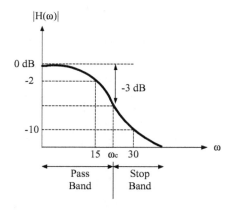

FIGURE 16.2 Low-pass Butterworth filter response of Example 16.1.

From Table 16.1

$$B(s) = s^2 + \sqrt{2}s + 1$$

$$H(s) = \frac{1}{B(s)}\Bigg|_{s=\frac{s}{\omega_c}}$$

$$H(s) = \frac{1}{s^2 + \sqrt{2}s + 1}\Bigg|_{s=\frac{s}{\omega_c}}$$

$$\omega_c = \frac{15}{\left(10^{\frac{2}{10}} - 1\right)^{1/2n}} = 17.2\,\mathrm{rad/sec}$$

$$H(s) = \frac{1}{\left(\dfrac{s}{17.2}\right)^2 + \sqrt{2}\,\dfrac{s}{17.2} + 1}$$

b) high-pass filter

Step 1: From the characteristics of the filter in the frequency domain, the order of the filter is calculated from

$$n = \frac{\log_{10}\left[\dfrac{10^{-\frac{k_1}{10}} - 1}{10^{-\frac{k_2}{10}} - 1}\right]}{2\log_{10}\left[\dfrac{\omega_2}{\omega_1}\right]} \tag{16.4}$$

Step 2: From the table, find how the function $B(s)$ is related to the order n:

n	$B(s)$
1	$S + 1$
2	$s^2 + \sqrt{2}s + 1$
3	$(s+1)(s^2 + s + 1)$
4	$(s^2 + 0.7653s + 1)(s^2 + 1.8477s + 1)$

Step 3: Find the cut-off frequency from Figure 16.3.

$$\omega_c = \omega_2 \tag{16.5}$$

Step 4:

$$H(s) = \frac{1}{B(s)}\bigg|_{s = \frac{\omega_c}{s}} \tag{16.6}$$

Example 16.2

Design an analog HPF that has pass-band attenuation of 3 dB at 5 rad/sec and stop-band attenuation of 15 dB at 2 rad/sec. Use Figure 16.4.

SOLUTION

$$n = \frac{\log_{10}\left[\dfrac{10^{-\frac{k_1}{10}} - 1}{10^{-\frac{k_2}{10}} - 1}\right]}{2\log_{10}\left[\dfrac{\omega_1}{\omega_2}\right]} = 2$$

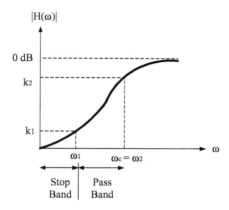

FIGURE 16.3 High-pass Butterworth filter response.

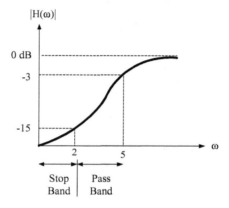

FIGURE 16.4 High-pass Butterworth filter response of Example 16.2.

From the table

$$B(s) = s^2 + \sqrt{2}s + 1$$

$$H(s) = \left.\frac{1}{B(s)}\right|_{s=\frac{\omega_c}{s}}$$

$$H(s) = \left.\frac{1}{s^2 + \sqrt{2}s + 1}\right|_{s=\frac{\omega_c}{s}}$$

$$H(s) = \frac{1}{\left(\dfrac{5}{s}\right)^2 + \sqrt{2}\,\dfrac{5}{s} + 1}$$

c) band-pass filter

Figure 16.5 shows the band-pass Butterworth filter response; the steps for design are as follows:

Step 1: From the characteristics of the filter in the frequency domain, the order of the filter is calculated from

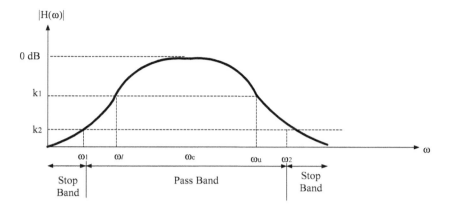

FIGURE 16.5 Band-pass Butterworth filter response.

$$n = \frac{\log_{10}\left[\dfrac{10^{-\frac{k_1}{10}}-1}{10^{-\frac{k_2}{10}}-1}\right]}{2\log_{10}[p]} \tag{16.7}$$

$$p = \frac{1}{\omega_r} \tag{16.8}$$

$$\omega_r = \min\left[|A| \text{ or } |B|\right] \tag{16.9}$$

$$|A| = \frac{-\omega_1^2 + \omega_u \omega_l}{\omega_1\left(\omega_u - \omega_l\right)} \tag{16.10}$$

$$|B| = \frac{\omega_2^2 - \omega_u \omega_l}{\omega_2\left(\omega_u - \omega_l\right)} \tag{16.11}$$

Step 2: From the table, find how the function $B(s)$ is related to the order n:

n	B(s)
1	$S+1$
2	$s^2 + \sqrt{2}s + 1$
3	$(s+1)(s^2+s+1)$
4	$(s^2 + 0.7653s + 1)(s^2 + 1.8477s + 1)$

Step 3: To find the transfer function in the s-plane

$$H(s) = \frac{1}{B(s)}\Bigg|_{s = \frac{s^2 - \omega_u \omega_l}{s(\omega_u - \omega_l)}} \tag{16.12}$$

2) Chebyshev filter

a) low-pass filter

Figure 16.6 shows the low-pass Chebyshev filter response; the steps for design are as follows:

Step 1:

$$k_1 = -10\log_{10}\left(1+\epsilon^2\right) \tag{16.13}$$

$$k_2 = -20\log_{10}(A) \tag{16.14}$$

$$g = \sqrt{\frac{A^2-1}{\epsilon^2}} \tag{16.15}$$

$$\omega_r = \frac{\omega_2}{\omega_1} \tag{16.16}$$

Step 2:

$$n = \frac{\log_{10}\left[g+\sqrt{g^2-1}\right]}{\log_{10}\left[\omega_r+\sqrt{\omega_r^2-1}\right]} \tag{16.17}$$

Step 3: From the table find the coefficients $b0$, $b1$, $b2$,…

$$H(s) = \frac{k_n}{V_n(s)} \tag{16.18}$$

$$k_n = \begin{cases} \dfrac{b_0}{\sqrt{1+\epsilon^2}} & n \text{ even} \\ b_0 & n \text{ odd} \end{cases} \tag{16.19}$$

$$V_n(s) = b_0 + b_1 s + b_2 s^2 + b_3 s^3 + \cdots + b_n s^n \tag{16.20}$$

$$H(s) = \frac{k_n}{V_n(s)}\bigg|_{s=\frac{s}{\omega_c}} \tag{16.21}$$

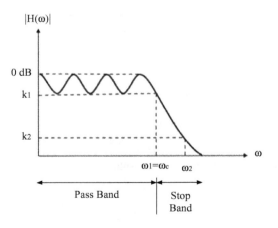

FIGURE 16.6   Low-pass Chebyshev filter response.

Example 16.3

Design a Chebyshev LPF for Figure 16.7.

SOLUTION

$$k_1 = -10\log_{10}\left(1+\epsilon^2\right)$$

$$-2 = -10\log_{10}\left(1+\epsilon^2\right)$$

$$\epsilon = 0.7647$$

$$k_2 = -20\log_{10}(A)$$

$$-20 = -20\log_{10}(A)$$

$A = 10$

$$g = \sqrt{\frac{A^2-1}{\epsilon^2}} = 13$$

$$\omega_r = \frac{\omega_2}{\omega_1} = 1.3$$

$$n = \frac{\log_{10}\left[g+\sqrt{g^2-1}\right]}{\log_{10}\left[\omega_r+\sqrt{\omega_r^2-1}\right]} = 4.3 \cong 5$$

$N = 5$ odd so $k_n = b_0$
From the table

$$V_n(s) = b_0 + b_1 s + b_2 s^2 + b_3 s^3 + b_4 s^4 + b_5 s^5$$

$$V_n(s) = 0.1228 + 0.5805s + 0.97439s^2 + 1.6888s^3 + 0.9368s^4 + 1s^5$$

$$H(s) = \left.\frac{k_n}{V_n(s)}\right|_{s=\frac{s}{\omega_c}}$$

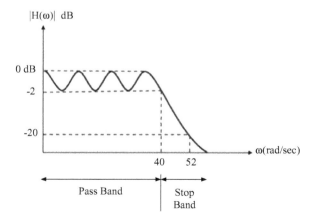

FIGURE 16.7 Low-pass Chebyshev filter response for Example 16.3.

$$H(s) = \frac{k_n}{0.1228 + 0.5805s + 0.97439s^2 + 1.6888s^3 + 0.9368s^4 + 1s^5}\Bigg|_{s=\frac{s}{\omega_c}}$$

$$H(s) = \frac{k_n}{0.1228 + 0.5805\dfrac{s}{40} + 0.97439\left(\dfrac{s}{40}\right)^2 + 1.6888\left(\dfrac{s}{40}\right)^3 + 0.9368\left(\dfrac{s}{40}\right)^4 + 1\left(\dfrac{s}{40}\right)^5}$$

b) high-pass filter

Figure 16.8 shows the high-pass Chebyshev filter response; the steps for design are as follows:

Step 1:

$$k_1 = -20\log_{10}(A) \tag{16.22}$$

$$k_2 = -10\log_{10}\left(1+\epsilon^2\right) \tag{16.23}$$

$$g = \sqrt{\frac{A^2-1}{\epsilon^2}} \tag{16.24}$$

$$\omega_r = \frac{\omega_2}{\omega_1} \tag{16.25}$$

Step 2:

$$n = \frac{\log_{10}\left[g+\sqrt{g^2-1}\right]}{\log_{10}\left[\omega_r+\sqrt{\omega_r^2-1}\right]} \tag{16.26}$$

Step 3: From the table find the coefficients $b0, b1, b2,\ldots$

$$H(s) = \frac{k_n}{V_n(s)}\Bigg|_{s=\frac{\omega_c}{s}} \tag{16.27}$$

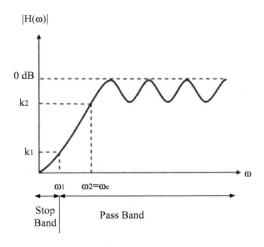

FIGURE 16.8 High-pass Chebyshev filter response.

$$k_n = \begin{cases} \dfrac{b_0}{\sqrt{1+\epsilon^2}} & n \text{ even} \\[2mm] b_0 & n \text{ odd} \end{cases} \tag{16.28}$$

$$V_n(s) = b_0 + b_1 s + b_2 s^2 + b_3 s^3 + \cdots + b_n s^n \tag{16.29}$$

$$H(s) = \left. \frac{k_n}{V_n(s)} \right|_{s = \frac{\omega_c}{s}} \tag{16.30}$$

Example 16.4

Design a Chebyshev HPF for the response shown in Figure 16.9, with a cut-off frequency of 40 rad/sec with a permissible ripple frequency of 2 rad/sec and stop-band attenuation of 20 dB at 15 rad/sec.

SOLUTION

$$k_2 = -10\log_{10}\left(1+\epsilon^2\right)$$

$$-2 = -10\log_{10}\left(1+\epsilon^2\right)$$

$$\epsilon = 0.7647$$

$$k_1 = -20\log_{10}(A)$$

$$-20 = -20\log_{10}(A)$$

$A = 10$

$$g = \sqrt{\frac{A^2-1}{\epsilon^2}} = 13$$

$$\omega_r = \frac{\omega_2}{\omega_1} = 2.6$$

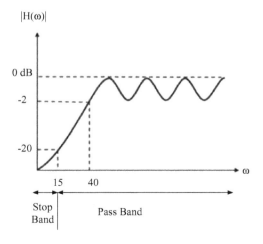

FIGURE 16.9   High-pass Chebyshev filter response for Example 16.4.

$$n = \frac{\log_{10}\left[g + \sqrt{g^2 - 1}\right]}{\log_{10}\left[\omega_r + \sqrt{\omega_r^2 - 1}\right]} = 2$$

$N = 2$ even so $k_n = b_0$
   From the table

$$V_n(s) = b_0 + b_1 s + b_2 s^2$$

$$V_n(s) = 0.63 + 0.8s + 1s^2$$

$$H(s) = \left.\frac{k_n}{V_n(s)}\right|_{s = \frac{\omega_c}{s}}$$

$$k_n = \frac{b_0}{\sqrt{1 + \epsilon^2}}$$

$$H(s) = \left.\frac{\dfrac{b_0}{\sqrt{1 + \epsilon^2}}}{0.63 + 0.8s + 1s^2}\right|_{s = \frac{s}{\omega_c}}$$

$$H(s) = \frac{k_n}{0.63 + 0.8\dfrac{40}{s} + 1\left(\dfrac{40}{s}\right)^2}$$

c) band-pass filter

Figure 16.10 shows the band-pass Chebyshev filter response; the steps for design are as follows:

Step 1: From the characteristics of the filter in the frequency domain, the order of the filter is calculated from

$$\omega_r = \min\left[\,|A|\text{ or }|B|\,\right] \tag{16.31}$$

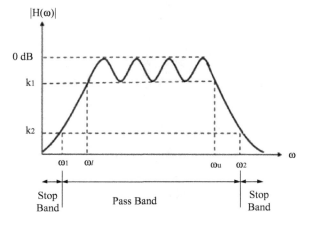

FIGURE 16.10   Band-pass Chebyshev filter response.

$$|A| = \frac{-\omega_1^2 + \omega_u \omega_l}{\omega_1 (\omega_u - \omega_l)}$$ (16.32)

$$|B| = \frac{\omega_2^2 - \omega_u \omega_l}{\omega_2 (\omega_u - \omega_l)}$$ (16.33)

Step 2:

$$k_1 = -10 \log_{10}(A)$$ (16.34)

$$k_2 = -20 \log_{10} \left(1 + \epsilon^2\right)$$ (16.35)

$$g = \sqrt{\frac{A^2 - 1}{\epsilon^2}}$$ (16.36)

$$n = \frac{\log_{10} \left[g + \sqrt{g^2 - 1} \right]}{\log_{10} \left[\omega_r + \sqrt{\omega_r^2 - 1} \right]}$$ (16.37)

Step 3: From the table, find how the function $B(s)$ is related to the order n:

n	$B(s)$
1	$S + 1$
2	$s^2 + \sqrt{2}s + 1$
3	$(s+1)(s^2 + s + 1)$
4	$(s^2 + 0.7653s + 1)(s^2 + 1.8477s + 1)$

Step 4: To find the transfer function in the s-plane

$$H(s) = \frac{1}{B(s)} \Bigg|_{s = \frac{s^2 + \omega_u \omega_l}{s(\omega_u - \omega_l)}}$$ (16.38)

16.1.2 BILINEAR TRANSFORMATION (IIR DIGITAL FILTER)

To design an IIR digital filter, a suitable transformation called a bilinear transformation is used; this transformation transforms the left-hand-side of the s-plane into the interior inside of the unit circle in the z-plane.

This bilinear transformation transforms the transfer function $H(s)$ of the analog filter

$$s = 2f_s \left(\frac{1 - z^{-1}}{1 + z^{-1}} \right)$$ (16.39)

f_s: sampling frequency.
Z^{-1}: delay element by T_s
if $s = j\omega$, $z^{-1} = e^{-j\lambda}$
where λ Digital frequency (rad)
ω: Analog frequency (rad/sec)

and the relation between λ and ω are given by

$$s = j\omega = \frac{2}{T_s}\left(\frac{1-e^{-j\lambda}}{1+e^{-j\lambda}}\right) \tag{16.40}$$

$$= \frac{2}{T_s}\left(\frac{e^{j\lambda/2} - e^{-j\lambda/2}}{e^{j\lambda/2} + e^{-j\lambda/2}}\right) \tag{16.41}$$

$$= \frac{2}{T_s}\tan\frac{\lambda}{2} \tag{16.42}$$

or

$$\lambda = 2\tan^{-1}\left(\frac{\omega T_s}{2}\right) \tag{16.43}$$

Example 16.5

Using a bilinear transformation design, realize a digital LPF with the following characteristics

   a Monotone pass-band
   b 3 dB cut-off frequency of $\pi/2$ rad
   c Stop-band attenuation of 15 dB at $3\pi/4$ rad($T_s = 1$ sec) (see Figure 16.11)

SOLUTION

Convert a digital frequency λ in rad to an analog frequency ω in rad/sec

$$\omega = \frac{2}{T_s}\tan\lambda/2$$

$$\omega_1 = \frac{2}{1}\tan\frac{\pi}{4} = 2\text{ rad/sec}$$

$$\omega_2 = \frac{2}{1}\tan 3\frac{\pi}{8} = 4.8\text{ rad/sec}$$

Figure 16.12 shows the low-pass bilinear transformation filter in terms of λ.

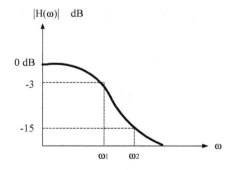

FIGURE 16.11   Low-pass bilinear transformation filter frequency response of Example 16.5.

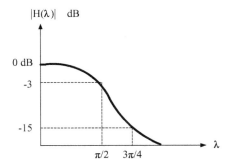

FIGURE 16.12 Low-pass bilinear transformation filter λ response of Example 16.5.

Monotone to Butterworth

$$n = \frac{\log_{10}\left[\dfrac{10^{-\frac{k_1}{10}}-1}{10^{-\frac{k_2}{10}}-1}\right]}{2\log_{10}\left[\dfrac{\omega_1}{\omega_2}\right]} = 2$$

From the table

$$B(s) = s^2 + \sqrt{2}s + 1$$

$$H(s) = \left.\frac{1}{B(s)}\right|_{s=\frac{s}{\omega_c}}$$

$$H(s) = \left.\frac{1}{s^2 + \sqrt{2}s + 1}\right|_{s=\frac{s}{\omega_c}}$$

$$\omega_c = \frac{\omega_1}{\left(10^{-\frac{k_1}{10}}-1\right)^{1/2n}} = 2 \text{ rad/sec}$$

$$H(s) = \frac{1}{\left(\dfrac{s}{2}\right)^2 + \sqrt{2}\dfrac{s}{2} + 1}$$

Design digital filter $T_s = 1$ sec so $f_s = 1$ Hz

$$s = 2f_s\left(\frac{1-z^{-1}}{1+z^{-1}}\right)$$

$$s = 2\left(\frac{1-z^{-1}}{1+z^{-1}}\right)$$

$$H(z) = \frac{1}{\left(\dfrac{1-z^{-1}}{1+z^{-1}}\right)^2 + \sqrt{2}\left(\dfrac{1-z^{-1}}{1+z^{-1}}\right) + 1}$$

$$H(z) = \frac{Y(z)}{X(z)} = \frac{1 + 2z^{-1} + z^{-2}}{3.4 + 0.58z^{-2}}$$

$$y(n) = \frac{1}{3.4}\left[x(n) + 2x(n-1) + x(n-2) - 0.58\, y(n-2) \right]$$

Figure 16.13 Shows the bilinear transformation of a LPF structure.

Example 16.6

Design of a second-order IIR low-pass digital filter using the bilinear transform method operates at a cut-off frequency of $\Omega_C = \pi/4$ radians/sample. The analog transfer function $H_a(s)$ for a second-order Butterworth low-pass filter with 3 dB cut-off at $\omega_C = 2\tan(\Omega_C/2) = 2 \tan(\pi/8)$ radians/second. Therefore, $\omega_C = 2 \tan(\pi/8) = 0.828$. It is well known by analog-filter designers that the transfer function for a second-order Butterworth low-pass filter with cut-off frequency $\omega = 1$ radian/second is:

$$H_a(s) = \frac{1}{1 + \left(\sqrt{2}\right)s + s^2}$$

When the cut-off frequency is $\omega = \omega_C$ rather than $\omega = 1$, the second-order expression for $H(s)$ becomes:

$$H_a(s) = \frac{1}{1 + \sqrt{2}(s/\omega_C) + (s/\omega_C)^2}$$

Replacing s with $j\omega$ and taking the modulus of this expression gives $G(\omega) = 1/\sqrt{[1 + (\omega/\omega_C)^{2n}]}$ with $n = 2$, which is a second-order Butterworth low-pass gain-response approximation. Deriving the above expression for $H_a(s)$ and corresponding expressions for higher orders is not part of our syllabus. It will not be necessary since MATLAB will take care of it.

Setting $\omega_C = 0.828$ in this formula, then replacing s by $2(z - 1)/(z + 1)$ gives us $H(z)$ for the required IIR digital filter.

Example 16.7

Design a second-order Butterworth IIR low-pass filter using the MATLAB program with a cut-off frequency of $\Omega_c = \pi/8$.

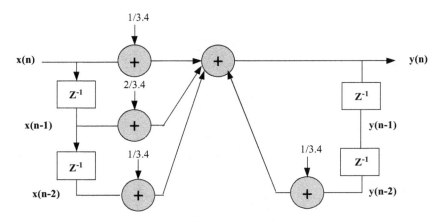

FIGURE 16.13 bilinear transformation LPF structure of Example 16.5.

SOLUTION

```
clc;
clear all;
[a b] = butter (2, 0.125)
a = [0.0300 0.0599 0.0300]
b = [1.0000 -1.4542 0.5741]
```

The required expression for $H(z)$ is

$$H(z) = \frac{0.0300 + 0.0599z^{-1} + 0.0300z^{-2}}{1 - 1.4542z^{-1} + 0.5741z^{-2}}$$

$$H(z) = 0.03\left(\frac{1 + 2z^{-1} + z^{-2}}{1 - 0.94z^{-1} + 0.33z^{-2}}\right)$$

Which may be realized by the signal-flow graph in Figure 16.14. Note the saving of two multipliers by using a multiplier to scale the input by 0.03.

16.1.3 HIGHER-ORDER IIR DIGITAL FILTERS

When recursive filters are of an order higher than two, then they are highly sensitive to quantization error and overflow. Therefore, higher-order IIR filters are designed as cascades of bi-quadratic sections.

Example 16.8

Design a fourth-order Butterworth-type IIR low-pass digital filter with a 3 dB cut-off at one-sixteenth of the sampling frequency f_S.

SOLUTION

Relative cut-off frequency is $\pi/8$. The MATLAB command below produces the arrays a and b with the numerator and denominator coefficients for the fourth-order system function $H(z)$.

```
clc;
clear all;
[a b] = butter(4, 0.125)
```

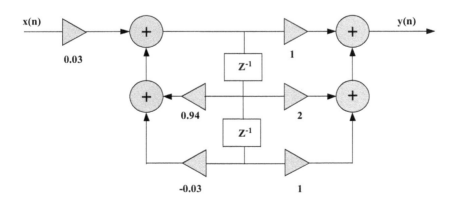

FIGURE 16.14 A second-order Butterworth IIR-LPF of Example 16.7.

The output produced by MATLAB is

```
a = 0.00093 0.0037 0.0056 0.0037 0.00093
b = 1.0000 -2.9768 3.4223 -1.7861 0.3556
```

The system function is, therefore, as follows:

$$H(z) = \left(\frac{0.00093 + 0.0037z^{-1} + .0056z^{-2} + .0037z^{-3} + 0.00093z^{-4}}{1 - 2.977z^{-1} + 3.422z^{-2} - 1.786z^{-3} + 0.3556z^{-4}} \right)$$

It corresponds to the fourth-order "direct-form" signal-flow graph shown in Figure 16.15.

Higher-order IIR digital filters are generally not implemented like this. They are implemented as cascaded bi-quad or second-order sections (SOS). Fortunately, MATLAB can transform the "direct-form" coefficients to second-order section (SOS) coefficients using a "Signal Processing Toolbox" function "tf2sos" as follows:

```
clc;
clear all;
[a b] = butter(4, 0.125)
[sos G] = tf2sos(a,b)
```

The output produced by MATLAB is

```
a = 0.0009 0.0037 0.0056 0.0037 0.0009
b = 1.0000 -2.9768 3.4223 -1.7861 0.3556
sos = 1.0000 2.0000 1.0000 1.0000 -1.3651 0.4776
      1.0000 2.0000 1.0000 1.0000 -1.6117 0.7445
G = 9.3350e-04
```

This produces a 2-dimensional array "SOS" containing two sets of bi-quad coefficients and a "gain" constant G. Mathematically, the system function based on this data is as follows:

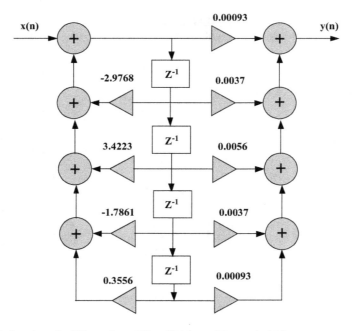

FIGURE 16.15   A fourth-order "direct-form II" realization of Example 16.8.

$$H(z) = 9.3350 \, \mathrm{e}^{-04} \left(\frac{1+2z^{-1}+z^{-2}}{1-1.365z^{-1}+0.478z^{-2}} \right) \left(\frac{1+2z^{-1}+z^{-2}}{1-1.612z^{-1}0.745z^{-2}} \right)$$

Practically, the effect of G is often distributed among the two sections, especially in fixed-point arithmetic, noting that, $0.033 \times 0.028 \approx 9.3350\mathrm{e}^{-04}$. Also, these two sections can be in order or an alternative expression for $H(z)$ is as follows:

$$H(z) = 0.033 \left(\frac{1+2z^{-1}+z^{-2}}{1-1.612z^{-1}0.745z^{-2}} \right) 0.028 \left(\frac{1+2z^{-1}+z^{-2}}{1-1.365z^{-1}+0.478z^{-2}} \right)$$

This alternative expression for $H(z)$ may be realized in the form of cascaded bi-quadratic sections, as shown in Figure 16.16.

16.1.4 IIR DIGITAL HIGH-PASS, BAND-PASS, AND BAND-STOP FILTER DESIGN

The bilinear transformation can apply to analog system functions, which are high-pass, band-pass, or band-stop to obtain digital filter equivalents. For example, a "high-pass" digital filter can be designed as illustrated below:

Example 16.8

Design a fourth-order high-pass IIR filter with cut-off frequency $fs/8$.

SOLUTION

Execute the following MATLAB commands and proceed as for a low-pass filter

```
clc;
clear all;
[a b] = butter(4,0.125,'high');
freqz(a,b);
[sos G] = tf2sos(a,b)
title('Response of HPF')
```

The output produced by MATLAB is:

```
a = 0.5963 -2.3852 3.5778 -2.3852 0.5963
b = 1.0000 -2.9768 3.4223 -1.7861 0.3556
sos = 1.0000 -2.0000 1.0000 1.0000 -1.3651 0.4776
      1.0000 -2.0000 1.0000 1.0000 -1.6117 0.7445
G = 0.5963
```

Figure 16.17 shows the Response of a fourth-order IIR-HPF using MATLAB.

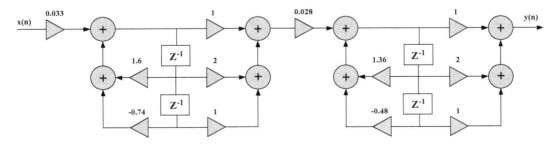

FIGURE 16.16 Fourth-order IIR Butterworth LPF with cut-off $fs/8$.

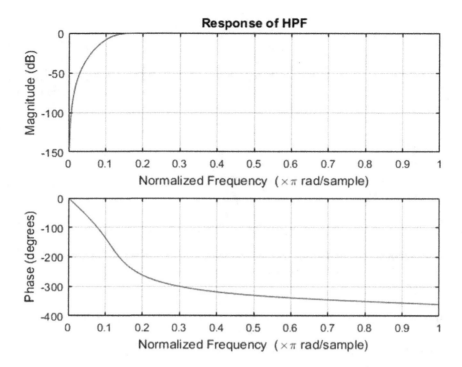

FIGURE 16.17   Response of fourth-order IIR-HPF of Example 16.8.

Example 16.9

Design a second(fourth)-order band-pass filter with $\omega_l = \pi/2$, $\omega_u = 3\pi/4$.

SOLUTION

Execute the following MATLAB statements:

```
clc;
clear all;
[a b] = butter(2,[0.5 0.75])
freqz(a,b);
[sos G] = tf2sos(a,b)
```

The MATLAB output is

```
a = 0.0976 0 -0.1953 0 0.0976
b = 1.0000 1.2190 1.3333 0.6667 0.3333
sos = 1.0000 -2.0000 1.0000 1.0000 0.1665 0.5348
      1.0000 2.0000 1.0000 1.0000 1.0524 0.6232
G = 0.0976
```

Figure 16.18 shows the response of a second(fourth)-order band-pass filter using the MATLAB program.

Example 16.10

Design a fourth(eighth)-order band-pass filter with $\omega_l = \pi/5$, $\omega_u = \pi/2.5$.

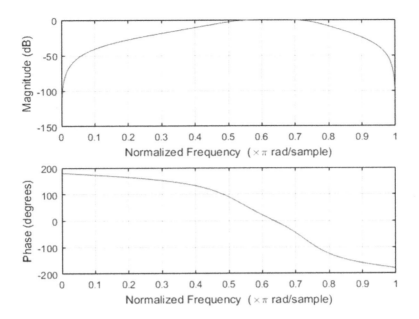

FIGURE 16.18   Response of second(fourth)-order band-pass filter for Example 16.9.

SOLUTION

Execute the following MATLAB statements:

```
clc;
clear all;
[a b] = butter(4,[0.2 0.4])
freqz(a,b); axis([0 1 -40 0]);
[sos G] = tf2sos(a,b)
```

The MATLAB output is

```
a = 0.0048 0 -0.0193 0 0.0289 0 -0.0193 0 0.0048
b =1.0000 -3.9366 8.2604 -11.2174 10.7836 -7.3914 3.5765 -1.1150 0.1874
sos =
 1.0000  -2.0000   1.0000  1.0000  -0.7574   0.5088
 1.0000   2.0001   1.0001  1.0000  -1.1268   0.5820
 1.0000   1.9999   0.9999  1.0000  -0.5824   0.7556
 1.0000  -2.0000   1.0000  1.0000  -1.4701   0.8374
G = 0.0048
```

Figure 16.19 response of second(eighth)-order band-pass filter using MATLAB program.

Example 16.11

Design a fourth(eighth)-order band-stop filter with $\omega_l = \pi/8$, $\omega_u = \pi/4$.

SOLUTION

Execute the following MATLAB statements:

```
clc;
clear all;
```

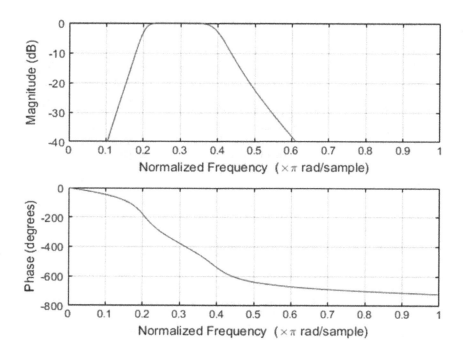

FIGURE 16.19   Response of second(eighth)-order band-pass filter for Example 16.10.

```
[a b] = butter(4,[0.125 0.25], 'stop')
freqz(a,b);
axis([0 1 -40 0]);
[sos G] = tf2sos(a,b)
```

The MATLAB output is

```
a = 0.5963 -4.0442 12.6706 -23.7586 29.0767 -23.7586 12.6706 -4.0442
0.5963
b = 1.0000 -5.9147 16.1669 -26.5104 28.4581 -20.4605 9.6300 -2.7200
0.3556
sos =
 1.0000 -1.6960 1.0006 1.0000 -1.3252 0.6585
 1.0000 -1.6950 0.9994 1.0000 -1.5276 0.7253
 1.0000 -1.6953 1.0001 1.0000 -1.3139 0.8278
 1.0000 -1.6958 0.9999 1.0000 -1.7480 0.8993
G =0.5963
```

Figure 16.20 response of fourth(eighth)-order band-stop filter using MATLAB program.

16.1.5   DESIGN A IIR LOW-PASS FILTER USING MATLAB

```
clc;
clear all;
fs =4000;
fc = 1500;
C = 2-cos(2*pi*fc/fs);
A(2) = sqrt(C^2 - 1)-C;
A(1) = 1;
```

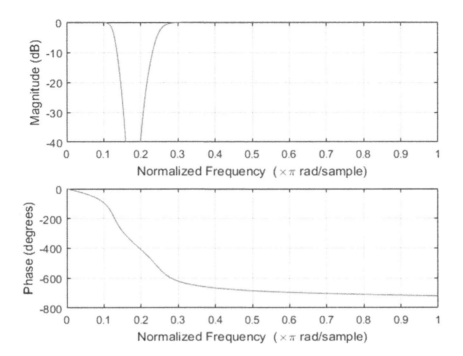

FIGURE 16.20 Response of fourth(eighth)-order band-stop filter for Example 16.11.

```
B = 1+A(2);
[H, w] = freqz(B, A);
H = 20*log10(abs(H));
plot(w/(2*pi)*fs/1000, H, fc/1000, -3, 'o');
title(' IIR-LPF Response ');
xlabel('Frequency (kHz)');
ylabel('Magnitude (dB)');
```

The output produced by MATLAB is shown in Figure 16.21.

16.1.6 DESIGN A IIR HIGH-PASS FILTER USING MATLAB

```
clc;
clear all;
fs = 20000;
fc = 1000;
C = 2-cos(2*pi*fc/fs);
A(2) = sqrt(C^2 - 1);
A(1) = 1;
B = 1-A(2);
[H, w] = freqz(B, A);
H = 20*log10(abs(H));
plot(w/(2*pi)*fs/1000, H);
title('IIR-HPF Response');
xlabel('Frequency (k Hz)');
ylabel('Magnitude (dB)');
```

The output produced by MATLAB is shown in Figure 16.22.

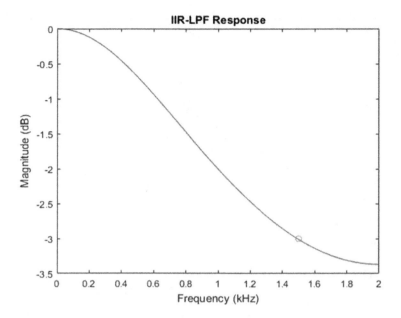

FIGURE 16.21   IIR-LPF response.

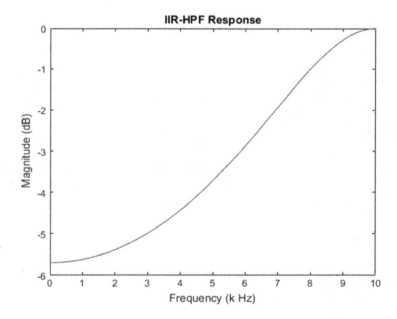

FIGURE 16.22   IIR high-pass filter response.

16.1.7   DESIGN AN IIR BAND-PASS FILTER USING MATLAB

```
fs =25000;
fc = 7000;
Bw = 4000;
A(3) = exp(-2*pi*(Bw/fs));
A(2) = -4*A(3)/(1+A(3))*cos(2*pi*fc/fs);
A(1) = 1;
```

```
B = 1-A(3) *sqrt(1-A(2)*A(2)/(4*A(3)));
[H, w] = freqz(B, A);
H = 20*log10(abs(H));
plot(w/(2*pi)*fs/1000, H);
title('IIR BPF Response');
xlabel('Frequency (k Hz)');
ylabel('Magnitude (dB)');
```

The output produced by MATLAB is shown in Figure 16.23.

16.2   FIR-FILTER DESIGN

For a certain FIR filter to be a linear phase, the following equation must be satisfied

$$h(n) = h(N-1-n) \quad n = 0,1,1...N-1$$

and N is the odd number of the coefficient.

1   Convert the unit of ω from rad/sec to rad.

$$\omega_{1\text{new}} = 2\pi f_L T_s \tag{16.44}$$

$$\omega_{2\text{new}} = 2\pi f_H T_s \tag{16.45}$$

$$\omega_{1\text{new}} = \omega_c \tag{16.45}$$

2   Calculate the number of samples N which must be an odd number

$$N = \frac{2\pi k}{\omega_{2\text{new}} - \omega_{1\text{new}}} \tag{16.46}$$

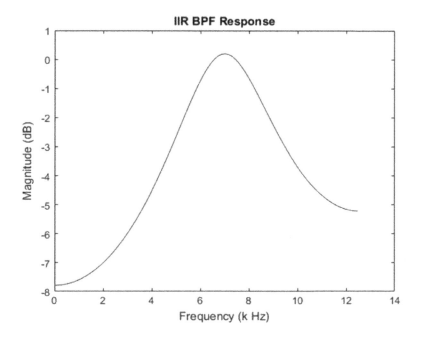

FIGURE 16.23   IIR BPF response.

3 Calculate

$$a = \frac{N-1}{2} \tag{16.47}$$

4 Using the type of window and type of filter.

16.2.1 DESIGN OF FIR FILTERS USING WINDOWS

The most straightforward approach for designing a FIR filter is to obtain a finite-length impulse response by truncating an infinite duration impulse response sequence. If $H_d\left(e^{j\omega}\right)$ is an ideal desired frequency response, then

$$H_d\left(e^{j\omega}\right) = \sum_{n=-\infty}^{\infty} h_d(n)e^{-jn\omega} \tag{16.48}$$

where $h_d(n)$ is the corresponding impulse response sequence, i.e.,

$$h_d(n) = \frac{1}{2\pi} \int_{-\pi}^{\pi} H_d\left(e^{j\omega}\right)e^{jn\omega}\,d\omega \tag{16.49}$$

The sequence $h_d(n)$, is of infinite duration, and it must be truncated to obtain a finite-duration impulse response. The window method is the one way to get a finite-duration causal impulse response is to known $h(n)$, and it is defined as

$$h(n) = \begin{cases} h_d(n) & 0 \le n \le N-1 \\ 0 & \text{otherwise} \end{cases} \tag{16.50}$$

Generally, we can represent $h(n)$ as the product of the desired impulse response and a finite duration "window" $w(n)$; i.e.,

$$h(n) = h_d(n)w(n) \tag{16.51}$$

where in the example

$$w(n) = \begin{cases} 1 & 0 \le n \le N-1 \\ 0 & \text{otherwise} \end{cases} \tag{16.52}$$

$$w\left(e^{jw}\right) = \sum_{n=0}^{N-1} e^{-jwn} = \frac{1-e^{-jwN}}{1-e^{-jw}} = e^{-jw\left(\frac{N-1}{2}\right)} \frac{\sin\left(\frac{wN}{2}\right)}{\sin\left(\frac{w}{2}\right)} \tag{16.53}$$

However, for the rectangular window, the "side lobes" are not insignificant. In fact, as N increases, the peak amplitudes of the main lobe and the side lobes grow in a manner such that the area under each lobe is a constant, while the width of each lobe decreases with N.

The main disadvantage of FIR filters is the possible requirement of a much higher-order filter (more coefficients) than IIR filters with comparable performance. As a result, it incurs longer delays, and higher computational costs and memory requirements. Also, there is no direct conversion from the popular analog-filter design to the digital FIR filter. Optimization of FIR-filter models requires computer programs for intensive computations.

The phase and group-delay characteristics of FIR filters are generally better than those of IIR filters. A FIR filter with good phase characteristics is usually the right choice for applications where wave shape is essential. However, IIR filters may be better than FIR filters for narrowband, sharp cut-off filters where phase is not necessary.

The I/O equation for a FIR filter of length L (or order $L - 1$) can be expressed as

$$y(n) = b_0 x(x) + b_1 x(n-1) + \cdots + b_{L-1} x(n-L+1)$$

$$= \sum_{i=0}^{L-1} b_i x(n-i) \tag{16.54}$$

$$= b^T x(n),$$

where T denotes the transposition of the vector, b is the filter coefficient vector ($L \times 1$) defined as

$$b = \begin{bmatrix} b_0 \, b_1 \ldots b_{L-1} \end{bmatrix}^T \tag{16.55}$$

and the signal vector (array or buffer) at time n is defined as

$$x(n) = \begin{bmatrix} x(n) \, x(n-1) \ldots x(n-L+1) \end{bmatrix}^L \tag{16.56}$$

The design of FIR filters determines the coefficients $\{b_i, i = 0,1, \ldots , L-1\}$ needed to achieve the desired filter characteristics with as few coefficients as possible.

Equation (16.54) implies that the FIR-filtering is equivalent to an inner (or dot) product of vectors or a linear convolution of the two sequences. A FIR filter is a non-recursive filter, which generates its output by simply scaling previous input samples by filter coefficients and then summing the weighted inputs. The coefficients are constants and determined by using the filter-design process. However, the signal vector $x(n)$ is a time function which needs to be updated every sampling period when the newest sample $x(n)$ is available.

By taking the z-transform of both sides of Equation (16.54) and using the time-shift property, we have

$$y(z) = b_0 x(z) + b_1 z^{-1} x(z) + \cdots b_{L-1} z^{-L+1} x(z) = x(z) \sum_{i=0}^{L-1} b_i z^{-i}$$

Digital filter design involves the computation of filter coefficients to approximate the desired frequency response. The design process starts from the filter specification and the implementation of the desired filter in the following five steps:

Step 1: Specification – The desired frequency response is determined by specifying filter characteristics in a frequency domain.

Step 2: Design criteria – A measure of the quality of the filter chosen by selecting design criteria.

Step 3: Realization – A class of filter chosen by selecting the filter type, structure, and length of the filter. An algorithm is then chosen to calculate the desired filter coefficients.

Step 4: Quantization – Quantization involves quantizing filter coefficients for a given word length, analyzing the finite-precision effects, and optimizing the quantized filter to reduce quantization errors.

Step 5: Implementation – Implementation involves verifying the designed filter using simulation and implementing it in the software (or hardware).

Some FIR-filter design methods may be simple such as the Fourier technique, which is introduced later. However, no single approach is optimal for every filter characteristic. Filter coefficients can be generated using some rational equations. However, using an optimization technique to obtain an optimum filter demands intensive computation. Also, the primary concern for implementing filters on fixed-point DSP processors for real-time applications is the limited precision. The theory for analyzing finite, word-length effects is too complicated to compute by hand; thus, we usually rely on computer software for designing digital filters.

A filter-design software package, such as MATLAB with the Signal Processing Toolbox and Filter-Design Toolbox, not only saves time when designing a filter quickly but also supports the design of a quantized filter for a given word length. We can explore many possibilities such as filter characteristics, structures, and different word lengths quickly using the software package.

Example 16.12

By the impulse response truncation method (via the windowing method during the rectangular window application) design a low-pass filter of order $N = 15$ with pass-band cut-off frequency (pass-band edge frequency) $f_0 = 1$ kHz. The sampling frequency is $f_s = 4$ kHz.

SOLUTION

$$f_S = 4\,\text{kHz}$$

$$f_0 = 1\,\text{kHz}$$

$$\omega_0 = \frac{2\pi}{f_s} f_o = \frac{2\pi}{4.10^3} \cdot 1.10^3 = \frac{\pi}{2}$$

$$H(e^{j\omega}) = \begin{cases} 1 & \text{for } |\omega| \le \omega_0 = \dfrac{\pi}{2} \\ 0 & \text{for } |\omega| > \omega_0 = \dfrac{\pi}{2} \end{cases}$$

$$h(n) = \frac{1}{2\pi} \int_{-\pi}^{\pi} H(e^{j\omega}) e^{j\omega n}\, d\omega = \frac{1}{2\pi} \int_{-\pi/2}^{\pi/2} 1 e^{j\omega n}\, d\omega$$

$$= \frac{1}{2\pi}\left[\frac{e^{j\omega n}}{jn}\right]_{-\pi/2}^{\pi/2} = \frac{1}{\pi n}\left[\frac{e^{jn\frac{\pi}{2}}}{2j} - \frac{e^{-jn\frac{\pi}{2}}}{2j}\right]$$

$$= \frac{1}{\pi n}\frac{e^{jn\frac{\pi}{2}} - e^{-jn\frac{\pi}{2}}}{2j} = \frac{\sin\left(n\dfrac{\pi}{2}\right)}{\pi n}$$

Then $h(0) = \mathbf{0.5}$

$$f(n) = h(n) \quad \text{for } -7 < n < 7$$

Example 16.13

Given the following digital system with a sampling rate of 8,000 Hz,

$$y(n) = x(n) - 0.5\, y(n-1)$$

Determine the frequency response of the system.

SOLUTION

Taking the z-transform of both sides of the difference equation, we get

$$Y(z) = X(z) - 0.5z^{-1}Y(z)$$

$$\Rightarrow Y(z) + 0.5z^{-1}Y(z) = X(z)$$

$$\Rightarrow \left(1 + 0.5z^{-1}\right)Y(z) = X(z)$$

Therefore, the transfer function of the system is given by

$$H(z) = \frac{Y(z)}{X(z)} = \frac{1}{\left(1 + 0.5z^{-1}\right)}$$

To find out the frequency response of the system, we replace z with $e^{j\Omega}$, which leads to

$$H\left(e^{j\Omega}\right) = \frac{1}{\left(1 + 0.5e^{-j\Omega}\right)}$$

It can be written as

$$H\left(e^{j\Omega}\right) = \frac{1}{\left(1 + 0.5\cos(\Omega) - j\,0.5\sin(\Omega)\right)}$$

Therefore, the magnitude frequency response and phase response are given by

$$\left|H\left(e^{\Omega}\right)\right| = \left|\frac{1}{\left(1 + 0.5\cos(\Omega) - j\,0.5\sin(\Omega)\right)}\right|$$

$$= \frac{1}{\sqrt{\left(1 + 0.5\cos(\Omega)\right)^2 + \left(-0.5\sin(\Omega)\right)^2}}$$

and

$$\angle H\left(e^{j\Omega}\right) = -\tan\left(\frac{-0.5\sin(\Omega)}{1 + 0.5\cos(\Omega)}\right)$$

This is an example of a high-pass filter.

Example 16.14

(a) Calculate the filter coefficients for a 5-tap FIR band-pass filter with a lower cut-off frequency of 2,000 Hz and an upper cut-off frequency of 2,400 at a sampling rate of 8,000 Hz.
(b) Determine the transfer function and the difference equation.
(c) Find out the frequency response of the filter.

SOLUTION

Part (a): We first determine the normalized cut-off frequencies

$$\Omega_L = \frac{2\pi f_L}{f_s} = 2\pi \times \frac{2000}{8000} = 0.5\pi \text{ radians}$$

$$\Omega_H = \frac{2\pi f_H}{f_s} = 2\pi \times \frac{2400}{8000} = 0.6\pi \text{ radians}$$

In this case $2M + 1 = 5$, therefore, from Table 15.1,

$$h(n) = \begin{cases} \dfrac{\Omega_H - \Omega_L}{\pi}, & n = 0 \\ \dfrac{\sin(\Omega_H\, n)}{n\pi} - \dfrac{\sin(\Omega_L\, n)}{n\pi}, & -2 \le n \le 2 \end{cases}$$

The non-causal FIR coefficients are

$$h(0) = \frac{\Omega_H - \Omega_L}{\pi} = \frac{0.6\pi - 0.5\pi}{\pi} = 0.1$$

$$h(1) = \frac{\sin(\Omega_H \times 1)}{\pi \times 1} - \frac{\sin(\Omega_L \times 1)}{\pi \times 1} = \frac{\sin(0.6\pi)}{\pi} - \frac{\sin(0.5\pi)}{\pi} = -0.01558$$

$$h(2) = \frac{\sin(\Omega_H \times 2)}{\pi \times 2} - \frac{\sin(\Omega_L \times 2)}{\pi \times 2} = \frac{\sin(1.2\pi)}{2\pi} - \frac{\sin(1.0\pi)}{2\pi} = -0.09355$$

Use the symmetry property

$$h(-1) = h(1) = -0.01558$$

$$h(-2) = h(2) = -0.09355$$

Thus, the filter coefficients are obtained by delaying by $M = 2$ samples, as

FILTER COEFFICIENTS

$$b_0 = h(0 - 2) = h(-2) = -0.09355$$

$$b_1 = h(1 - 2) = h(-1) = -0.01558$$

$$b_2 = h(2 - 2) = h(0) = 0.1$$

$$b_3 = h(3 - 2) = h(1) = -0.01558$$

$$b_4 = h(4 - 2) = h(2) = -0.09355$$

Part (b): Therefore, the transfer function in this case is

$$H(z) = b_0 + b_1 z^{-1} + b_2 z^{-2} + b_3 z^{-3} + b_4 z^{-4}$$

$$= -0.09355 - 0.01558 z^{-1} + 0.1 z^{-2}$$

$$- 0.01558 z^{-3} - 0.09355 z^{-4}$$

The difference equation is

$$y(n) = -0.09355\, x(n) - 0.01558\, x(n-1) + 0.1 x(n-2)$$

$$- 0.01558\, x(n-3) - 0.09355\, x(n-4)$$

Part (c): The frequency response of the filter is

$$H\left(e^{j\Omega}\right) = -0.09355 - 0.01558\,e^{-j\Omega} + 0.1e^{-2j\Omega}$$

$$-0.01558\,e^{-3j\Omega} - 0.09355\,e^{-4j\Omega}$$

Casual FIR-filter coefficients: shifting $h(n)$ to the right by samples 1, 2, …, M.
   T ransfer Function:

$$H(z) = b_0 + b_1z^{-1} + b_2z^{-2} + \cdots + b_{2M}z^{-2M}$$

where

$$b_n = h(n - M) \quad for \ n = 0,1,2,\ldots,2$$

Example 16.15

A digital filter with a sampling rate of 1,000 Hz is required to eliminate an unwanted 105 Hz sinusoidal component of an input signal without affecting the magnitudes of other parts too severely. Design a fourth-order "notch" filter for this purpose whose 3 dB bandwidth is not greater than 10 Hz.

SOLUTION

```
clc;
clear all;
FS=1000;
FL=100;
FU=110;
[a b]=butter(2,[FL,FU]/(FS/2), 'stop')
freqz(a,b);
title('Response of SPF')
[SOS G] = tf2sos(a,b)
```

The MATLAB output is

```
a = 0.9565 -3.0248 4.3043 -3.0248 0.9565
b = 1.0000 -3.0920 4.3024 -2.9575 0.9150
SOS = 1.0000 -1.5811 1.0000 1.0000 -1.5184 0.9553
      1.0000 -1.5811 1.0000 1.0000 -1.5736 0.9578
G = 0.9565
```

The output produced by MATLAB is shown in Figure 16.24.

Example 16.16

Design a second-order low-pass digital filter. Given:

$$H(s) = \frac{\omega_c^2}{s^2 + \sqrt{2}\omega_c s + \omega_c^2}$$

The sampling frequency is 5 Hz, and the desired low-pass cut-off frequency is 0.318 Hz (2 rad/sec). Our desired digital cut-off frequency is 2 rd/sec * (1/5 Hz) = 0.4 rad.

SOLUTION

$$\Omega_c = \frac{2}{T}\tan\left(\frac{\omega_c}{2}\right) = \frac{2}{(1/5)}\tan\left(\frac{0.4}{2}\right) = 2.027 \text{ rd/sec}$$

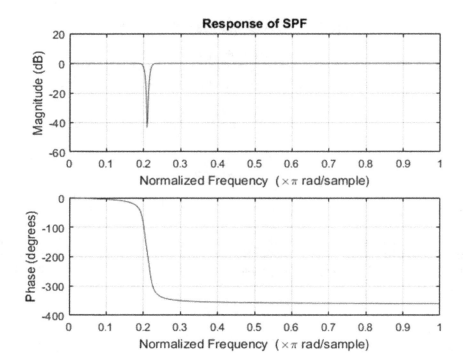

FIGURE 16.24 Response of a fourth-order "notch" filter of Example 16.15.

$$f_c = \Omega_c / 2\pi = 0.323 \, \text{Hz (shifted from 0.318 Hz)}$$

$$H(s) = \frac{\omega_c^2}{s^2 + \sqrt{2}\omega_c s + \omega_c^2} = \frac{0.16}{s^2 + 0.567s + 0.16}$$

$$H(z) = H(s)\big|_{s=\frac{2}{T}\frac{z-1}{z+1}} = \frac{0.0309(1 + 2z^{-1} + z^{-2})}{1 - 1.444z^{-1} + 0.5682z^{-2}}$$

PROBLEMS

16.1 For the digital filter of the z-plane pole and zero patters shown in Figure 16.25, find the filter transfer function, difference equation, and sketch the impulse response sequence with a filter gain of 10.

16.2 Use the bilinear transformation to design a digital filter based on a three-order Butterworth HPF $H(s) = 1/S(S^2 + 1.414S + 1)$. The digital filter should have a digital cut-off frequency equal to 10 kHz, while the sampling frequency of the input signal is 1 kHz.

16.3 Design the digital low-pass filter with a 3-dB cut-off frequency, $\omega_c = 0.25\pi$ using the bilinear transformation method to analog a Butterworth low-pass filter defined by

$$H_a(s) = [1/1 + (s / \Omega_s)]$$

16.4 Determine the low-pass digital filter with a 3-dB cut-off frequency of 0.2π and its frequency response from the analog filter given by $H_a(s) = \Omega_c/s + \Omega_c$, where Ω_c is the 3-dB cut-off frequency.

16.5 Design a LPF IIR digital filter with a 3 dB cut-off frequency of 1 kHz and stop-band attenuation of 28 dB at 2 kHz. Use $f_s = 8\text{kHz}$ and the Chebyshev approximation.

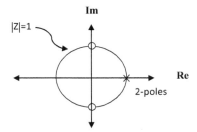

FIGURE 16.25 *z*-plain of Problem 16.1.

16.6 Design an HPF IIR digital filter with a monotone pass-band and a cut-off frequency of $\pi/2$ rad with attenuation of –3 dB stop-band attenuation of –15 dB at a frequency of $\pi/4$. Use $f_s =$ 1 Hz and the Chebyshev approximation.

Bibliography

1. Vijay K. Madisetti and Douglas B. Williams, *Digital Signal Processing Hand Book*, CRC net base, 1999.
2. D. S. G. Pollock, *A Handbook of Time-Series Analysis, Signal Processing and Dynamics, Signal Processing and Its Applications*, Academic Press, 1999.
3. Alan V. Oppenheim, *Signals and Systems*, Prentice-Hall Signal Processing Series, 1996.
4. Richard G. Lyons, *Understanding Digital Signal Processing*, Prentice Hall PTR, 1997.
5. Jonathan Y. Stein, *Digital Signal Processing: A Computer Science Perspective*, A Wiley-Interscience Publication, John Wiley & Sons, Inc., 2000.
6. John G. Proakis and Dimitris G. Manolakis, *Digital Signal Processing Principles, Algorithms, and Applications*, 3rd Edition, Prentice-Hall International, Inc., 1996.
7. Stanley H. Mneney, *An Introduction to Digital Signal Processing: A Focus on Implementation*, River Publishers ApS, 2008
8. Allen B. Downey, *Think DSP Digital Signal Processing in Python Version 1.0.9*, Green Tea Press, Needham, MA, 2014.
9. Vinay K. Ingle and John G. Proakis, *Digital Signal Processing Using MATLAB®*, 3rd Edition, Publisher, Global Engineering: Christopher M. Shortt, 2012.
10. Jonathan Blackledge, *Digital Signal Processing*, 2nd Edition, Horwood Publishing, Chichester, West Sussex, England, 2006.
11. Vijay K. Madisetti and Douglas B. Williams, *Digital Signal Processing Handbook*, CRC Press LLC, 1999.
12. Monson H. Hayes, *Schaum's Outline of Theory and Problems of Digital Signal Processing*, McGraw-Hill, 1998.
13. Andreas Antoniou, *Digital Signal Processing: SIGNALS SYSTEMS AND FILTERS*, McGraw-Hill, 2006.
14. Matthew N. O. Sadiku and Warsame Hassan Ali, *Signals and Systems: A Primer with MATLAB®*, 1st Edition, CRC Press, 2015.
15. Brigham, E. O., *The Fast Fourier Transform and Its Applications*, Prentice Hall, Englewood Cliffs, NJ, 1988.
16. Buck, J. R., M. M. Daniel, and A. C. Singer, *Computer Explorations in Signals and Systems*, 2nd Edition, Prentice Hall, Upper Saddle River, NJ, 2002.
17. Cadzow, J. A. and H. F. Van Landingham, *Signals, Systems, and Transforms*, Prentice Hall, Englewood Cliffs, NJ, 1985.
18. Carlson, G. E., *Signal and Linear System Analysis with MATLAB*, 2nd Edition, John Wiley & Sons, New York, 1998.
19. Chapiro, L. F., *Signals and Systems Using MATLAB*, Elsevier, Burlington, MA, 2011.
20. Chen, C., *Linear System Theory and Design*, 3rd Edition, Oxford University Press, New York, 1999.
21. Chen, C., *Signals and Systems*, 3rd Edition, Oxford University Press, New York, 2004.
22. Denbigh, P., *System Analysis and Signal Processing*, Addison Wesley Longman, Essex, UK, 1998.
23. ElAli, T. S., *Discrete Systems and Digital Signal Processing with MATLAB*, CRC Press, Boca Raton, FL, 2004.
24. Anderson, A. H. et al., VHDL executable requirements, *Proceedings 1st Annual RASSP Conference*, pp. 87–90, Arlington, VA, August, 1994. URL: http://rassp.scra.org/public/confs/1st/papers.html#VER.
25. Shaw, G. A. and Anderson, A. H., Executable requirements: Opportunities and impediments, in *IEEE Proceedings of the International Conference on Acoustics, Speech, and Signal Processing*, pp. 1232–1235, Atlanta, GA, May 7–10, 1996.
26. Frank, G. A., Armstrong, J. R., and Gray, F. G., Support for model-year upgrades in VHDL test benches, in *Proceedings of the 2nd Annual RASSP Conference*, pp. 211–215, Arlington, VA, July 24–27, 1995. URL: http://rassp.scra.org/public/confs/2nd/papers.html.
27. ISO/IEC 11172, *Information Technology—Coding of Moving Picture and Associated Audio for Digital Storage Media at up to about 1.5 Mbit/s*, 1993. https://www.iso.org/standard/22412.html
28. Rowe, L. A., Patel, K. et al., *mpeg encode/mpeg play, Version 1.0*, available via anonymous ftp at ftp://mm-ftp.cs.berkeley.edu/pub/multimedia/mpeg/bmt1r1.tar.gz, Computer Science Department, EECS University of California at Berkeley, May 1995.
29. International Standard ISO/IEC 13818, Information technology -Generic coding of moving pictures and associated audio information: Systems Second edition, 2000.
30. S. A. Tretter, *Communications System Design Using DSP Algorithms*, Plenum Press, New York, NY, 1995.

31. Egolf, T. W., Famorzadeh, S., and Madisetti, V. K., Fixed-point co-design in DSP, in *VLSI Signal Processing Workshop*, Vol. 8, October, 1994.

32. Thomas, D., Adams, J., and Schmit, H., A model and methodology for hardware-software codesign, in *IEEE Design & Test of Computers*, September, 1993.

33. Kalavade, A. and Lee, E., A global criticality/local phase driven algorithm for the constrained hardware/software partitioning problem, in *Proceedings of the Third International Workshop on Hardware/ Software Codesign*, September, 1994.

34. Bracewell, R. N., *The Fourier Transform and Its Applications*, 2nd Edition, McGraw-Hill, New York, 1978.

35. R. Schaumann and M. E. Van Valkenburg, *Design of Analog Filters*, New York: Oxford University Press, 2001.

36. Alan V. Oppenheim and Ronald W. Schafer, *Discrete-Time Signal Processing*, 2nd Edition, *Prentice Hall Signal Processing Series*, 1999, ISBN 0-13-754920-2.

37. Steven W. Smith, *The Scientist and Engineer's Guide to Digital Signal Processing*, California Technical Publishing, 2007.

38. A. Antoniou, *Digital Filters: Analysis, Design, and Applications*, New York, McGraw-Hill, 1993.

39. Dragoslav Mitronivić and Jovan, Kečkić, *The Cauchy Method of Residues: Theory and Applications*, D. Reidel Publishing Company, 1984, ISBN 90-277-1623-4.

40. Wolfram Mathworld, *Partial Fraction Decomposition*. URL: http://mathworld.wolfram.com/ PartialFractionDecomposition.html. Last accessed 10th May 2008.

41. J. Eyre and J. Bier, "The evolution of DSP Processors," *A BDTI White Paper*, Berkeley Design Technology Inc., 2000.

42. A. V. Oppenheim and R. W. Schafer, *Discrete-Time Signal Processing*, Prentice Hall, 1989.

43. A. V. Oppenheim and R. W. Schafer, *Digital Signal Processing*, Prentice Hall, 1975.

44. A. Papoulis, *Signal Analysis*, McGraw-Hill, 1977.

45. C. S. Burrus, J. H. McClellan, A. V. Oppenheim, T. W. Parks, R. W. Schafer, H. W. Schuessler, *Computer-Based Exercises for Signal Processing Using Matlab*, Prentice Hall, 1994.

46. L. R. Rabiner and R. W. Schafer, *Digital Processing of Speech*, Prentice Hall, 1978.

47. J. M. Mendel, *Lessons in Estimation Theory for Signal Processing, Communications and Control*, Prentice Hall, 1995.

48. D. B. Percival and A. T. Walden, *Spectral Analysis for Physical Applications*, Cambridge Press, 1993.

49. Candy, J. V. *Signal Processing*, McGraw-Hill, 1988.

50. S. K. Mitra, *Digital Signal Processing: A Computer Based Approach*, 2nd Edition, McGraw-Hill, 2001.

51. A. Antoniou, *Digital Filters: Analysis, Design and Applications*, 2nd Edition, Mc-Graw-Hill, New York, NY, 1993.

52. *DSP Selection Guide, SSDV004S*, Texas Instrument Inc., Printed by Southwest Precision Printers, Houston, Texas, 2007.

53. G. Marven and G. Ewers, *A Simple Approach to Digital Signal Processing*, Texas Instruments. ISBN 0-904 047-00-8, 1994.

54. Michael R. Williams, *A History of Computing Technology*, IEEE Computer Society Press, ISBN 0-8186-7739-2, 1997.

55. A. W. Burks, H. H. Goldstine, and J. von Neumann, *Preliminary Discussion of the Logical Design of an Electronic Computing Instrument*, 1963.

56. P. A. Regalia, *Adaptive IIR Filtering for Signal Processing Control*, Marcel Dekker, New York, 1995.

57. A. V. Oppenheim and R. W. Schafer, *Discrete-Time Signal Processing*, Englewood Cliffs, NJ: Prentice-Hall, 1989.

58. A. Papoulis, *Probability, Random Variables, and Stochastic Processes*, McGraw-Hill, New York, 1991.

59. D. S. G. Pollock, *A Handbook of Time-Series Analysis, Signal Processing and Dynamics, Signal Processing and Its Applications*, Academic Press, 1999.

Appendix A: Complex Numbers

The ability to handle complex numbers is important in signals and systems. Although calculators and computer software packages such as MATLAB are now available to manipulate complex numbers, it is advisable that students be familiar with how to handle them by hand.

A.1   REPRESENTATION OF COMPLEX NUMBERS

A complex number z may be written in *rectangular form* as

$$z = x + jy \tag{A.1}$$

where $j = \sqrt{-1}$; x is the real part of z while y is the imaginary part; that is

$$x = \text{Re}(z), \quad y = \text{Im}(z) \tag{A.2}$$

Since $j = \sqrt{-1}$

$$\frac{1}{j} = -j$$

$$j^2 = -1$$

$$j^3 = j \cdot j^2 = -j \tag{A.3}$$

$$j^4 = j^2 \cdot j^2 = 1$$

$$j^5 = j \cdot j^4 = j$$

$$\vdots$$

A second way of representing the complex number z is by specifying its magnitude r and angle θ it makes with the real axis, as shown in Figure B.1. This is known as the *polar form*. It is given by

$$z = |z| \angle \theta = r \angle \theta \tag{A.4}$$

where

$$r = \sqrt{x^2 + y^2}, \quad \theta = \tan^{-1} \frac{y}{x} \tag{A.5a}$$

or

$$x = r\cos\theta, \quad y = r\sin\theta \tag{A.5b}$$

that is,

$$z = x + jy = r\angle\theta = r\cos\theta + jr\sin\theta \tag{A.6}$$

In converting from rectangular to polar form using Equation (A.5), we must exercise care in determining the correct value of θ. These are the four possibilities:

$$z = x + jy, \quad \theta = \tan^{-1} \frac{y}{x} \quad \text{(1st quadrant)}$$

$$z = -x + jy, \quad \theta = 180° - \tan^{-1} \frac{y}{x} \quad \text{(2nd quadrant)}$$

$$z = -x - jy, \quad \theta = 180° + \tan^{-1} \frac{y}{x} \quad \text{(3rd quadrant)} \tag{A.7}$$

$$z = x - jy, \quad \theta = 360° - \tan^{-1} \frac{y}{x} \quad \text{(4th quadrant)}$$

if x and y are positive.

The third way of representing the complex number x is the *exponential form*:

$$z = re^{j\theta} \tag{A.8}$$

This is almost the same as the polar form, because we use the same magnitude r and the angle θ.

The three forms of representing a complex number are summarized as follows.

$$z = x + jy, \quad (x = r\cos\theta, y = r\sin\theta) \quad \text{Rectangular form}$$

$$z = r\angle\theta, \quad \left(r = \sqrt{x^2 + y^2}, \theta = \tan^{-1} \frac{y}{x}\right) \quad \text{Polar form} \tag{A.9}$$

$$z = re^{j\theta}, \quad \left(r = \sqrt{x^2 + y^2}, \theta = \tan^{-1} \frac{y}{x}\right) \quad \text{Exponential form}$$

A.2 MATHEMATICAL OPERATIONS

Two complex numbers $z_1 = x_1 + jy_1$ and $z_2 = x_2 + jy_2$ are equal if and only their real parts are equal and their imaginary parts are equal, that is

$$x_1 = x_2, \quad y_1 = y_2 \tag{A.10}$$

The complex conjugate of the complex number $z = x + jy$ is

$$z^* = x - jy = r\angle - \theta = re^{-j\theta} \tag{A.11}$$

Thus, the complex conjugate of a complex number is found by replacing every j by $-j$.

Given two complex number $z_1 = x_1 + jy_1 = r_1\angle\theta_1$ and $z_2 = x_2 + jy_2 = r_2\angle\theta_2$, their sum is

$$z_1 + z_2 = (x_1 + x_2) + j(y_1 + y_2) \tag{A.12}$$

and their difference is

$$z_1 - z_2 = (x_1 - x_2) + j(y_1 - y_2) \tag{A.13}$$

While it is more convenient to perform addition and subtraction of complex numbers in rectangular form, the product and quotient of two complex numbers are best done in polar or exponential form. For their product,

$$z_1 z_2 = r_1 r_2 \angle \theta_1 + \theta_2 \tag{A.14}$$

Alternatively, using the rectangular form

$$
\begin{aligned}
z_1 z_2 &= (x_1 + jy_1)(x_2 + jy_2) \\
&= (x_1 x_2 - y_1 y_2) + j(x_1 y_2 + x_2 y_1)
\end{aligned}
\tag{A.15}
$$

For their quotient,

$$
\frac{z_1}{z_2} = \frac{r_1}{r_2} \angle \theta_1 - \theta_2
\tag{A.16}
$$

Alternatively, using the rectangular form,

$$
\frac{z_1}{z_2} = \frac{x_1 + jy_1}{x_2 + jy_2}
\tag{A.17}
$$

We rationalize the denominator by multiplying both the numerator and denominator by z_2^*.

$$
\frac{z_1}{z_2} = \frac{(x_1 + jy_1)(x_2 - jy_2)}{(x_2 + jy_2)(x_2 - jy_2)} = \frac{x_1 x_2 + y_1 y_2}{x_2^2 + y_2^3} + j\,\frac{x_2 y_1 - x_1 y_2}{x_2^2 + y_2^3}
\tag{A.18}
$$

A.3 EULER'S FORMULA

Euler's formula is an important result in complex variables. We derive it from the series expansion of e^x, $\cos\theta$, and $\sin\theta$. We know that

$$
e^x = 1 + x + \frac{x^2}{2!} + \frac{x^3}{3!} + \frac{x^4}{4!} + \cdots
\tag{A.19}
$$

Replacing x by $j\theta$ gives

$$
e^{j\theta} = 1 + j\theta - \frac{\theta^2}{2!} - j\frac{\theta^3}{3!} + \frac{\theta^4}{4!} + \cdots
\tag{A.20}
$$

Also,

$$
\cos\theta = 1 - \frac{\theta^2}{2!} + \frac{\theta^4}{4!} - \frac{\theta^6}{6!} + \cdots
\tag{A.21a}
$$

$$
\sin\theta = \theta - \frac{\theta^3}{3!} + \frac{\theta^5}{5!} - \frac{\theta^7}{7!} + \cdots
\tag{A.21b}
$$

so that

$$
\cos\theta + j\sin\theta = 1 + j\theta - \frac{\theta^2}{2!} - j\frac{\theta^3}{3!} + \frac{\theta^4}{4!} + j\frac{\theta^5}{5!} - \cdots
\tag{A.22}
$$

Comparing Equations (A.20) and (A.22), we conclude that

$$
e^{j\theta} = \cos\theta + j\sin\theta
\tag{A.23}
$$

This is known as Euler's formula. The exponential form of representing a complex number as in Equation (A.8) is based on Euler's formula. From Equation (A.23), notice that

$$
\cos\theta = \mathrm{Re}(e^{j\theta}), \quad \sin\theta = \mathrm{Im}(e^{j\theta})
\tag{A.24}
$$

and that

$$\left| e^{j\theta} \right| = \sqrt{\cos^2 \theta + \sin^2 \theta} = 1 \tag{A.25}$$

Replacing θ by $-\theta$ in Equation (A.23) gives

$$e^{-j\theta} = \cos \theta - j \sin \theta \tag{A.26}$$

Adding Equations (A.23) and (A.26) yields

$$\cos \theta = \frac{1}{2} \left(e^{j\theta} + e^{-j\theta} \right) \tag{A.27}$$

Subtracting Equation (A.26) from Equation (A.23) yields

$$\sin \theta = \frac{1}{2j} \left(e^{j\theta} - e^{-j\theta} \right) \tag{A.28}$$

The following identities are useful in dealing with complex numbers. If $z = x + jy = r \angle \theta$, then

$$zz^* = |z|^2 = x^2 + y^2 = r^2 \tag{A.29}$$

$$\sqrt{z} = \sqrt{x + jy} = \sqrt{r} e^{j\theta/2} = \sqrt{r} \angle \theta / 2 \tag{A.30}$$

$$z^n = (x + jy)^n = r^n \angle n\theta = r^n (\cos n\theta + j \sin n\theta) \tag{A.31}$$

$$z^{1/n} = (x + jy)^{1/n} = r^{1/n} \angle \theta / n + 2\pi k / n, \quad k = 0, 1, 2, \ldots, n - 1 \tag{A.32}$$

$$\ln(re^{j\theta}) = \ln r + \ln e^{j\theta} = \ln r + j\theta + j2\pi k \quad (k = \text{integer}) \tag{A.33}$$

$$
\begin{aligned}
e^{\pm j\pi} &= -1 \\
e^{\pm j2\pi} &= 1 \\
e^{j\pi/2} &= j \\
e^{-j\pi/2} &= -j
\end{aligned}
\tag{A.34}
$$

$$
\begin{aligned}
\text{Re}\left(e^{(a + j\omega)t} \right) &= \text{Re}\left(e^{at} e^{j\omega t} \right) = e^{at} \cos \omega t \\
\text{Im}\left(e^{(a + j\omega)t} \right) &= \text{Im}\left(e^{at} e^{j\omega t} \right) = e^{at} \sin \omega t
\end{aligned}
\tag{A.35}
$$

Appendix B: Mathematical Formulas

The appendix contains all the formulas needed to solve the problems in this book.

B.1   QUADRATIC FORMULAS

The roots of the quadratic equation

$$ax^2 + bx + c = 0$$

$$x_1, x_2 = \frac{-b \pm \sqrt{b^2 - 4ac}}{2a}$$

B.2   TRIGONOMETRIC IDENTITIES

$$\sin(-x) = -\sin x$$

$$\cos(-x) = \cos x$$

$$\sec x = \frac{1}{\cos x}, \quad \csc x = \frac{1}{\sin x}$$

$$\tan x = \frac{\sin x}{\cos x}, \quad \cot x = \frac{1}{\tan x}$$

$$\sin(x \pm 90°) = \pm \cos x$$

$$\cos(x \pm 90°) = \mp \sin x$$

$$\sin(x \pm 180°) = -\sin x$$

$$\cos(x \pm 180°) = -\cos x$$

$$\cos^2 x + \sin^2 x = 1$$

$$\frac{a}{\sin A} = \frac{b}{\sin B} = \frac{c}{\sin C} \quad \text{(law of sines)}$$

$$a^2 = b^2 + c^2 - 2bc \cos A \quad \text{(law of cosines)}$$

$$\frac{\tan \frac{1}{2}(A - B)}{\tan \frac{1}{2}(A + B)} = \frac{a - b}{a + b} \quad \text{(law of tangents)}$$

$$\sin(x \pm y) = \sin x \cos y \pm \cos x \sin y$$

$$\cos(x \pm y) = \cos x \cos y \mp \sin x \sin y$$

$$\tan(x \pm y) = \frac{\tan x \pm \tan y}{1 \mp \tan x \tan y}$$

$$2 \sin x \sin y = \cos(x - y) - \cos(x + y)$$

$$2 \sin x \cos y = \sin(x + y) - \sin(x - y)$$

$$2 \cos x \cos y = \cos(x + y) - \cos(x - y)$$

$$\sin 2x = 2 \sin x \cos x$$

$$\cos 2x = \cos^2 x - \sin^2 x = 2 \cos^2 x - 1 = 1 - 2 \sin^2 x$$

$$\tan 2x = \frac{2 \tan x}{1 - \tan^2 x}$$

$$\cos^2 x = \frac{1 + \cos(2x)}{2}$$

$$\sin^2 x = \frac{1 - \cos(2x)}{2}$$

$$a \cos x + b \sin x = K \cos(x + \theta), \quad \text{where } K = \sqrt{a^2 + b^2} \quad \text{and } \theta = \tan^{-1}\left(\frac{-b}{a}\right)$$

$$e^{\pm jx} = \cos x \pm j \sin x \quad \text{(Euler's formula)}$$

$$\cos x = \frac{e^{jx} + e^{-jx}}{2}$$

$$\sin x = \frac{e^{jx} - e^{-jx}}{2j}$$

B.3   TRIGONOMETRIC SUBSTITUTION

Form	Trig Sub	Identity
$\sqrt{a^2 + x^2}$	$x = a \tan \theta$	$1 + \tan^2 \theta = \sec^2 \theta$
$\sqrt{a^2 - x^2}$	$x = a \sin \theta$	$1 - \sin^2 \theta = \cos^2 \theta$
$\sqrt{x^2 - a^2}$	$x = a \sec \theta$	$\sec^2 \theta - 1 = \tan^2 \theta$

B.4 HYPERBOLIC FUNCTIONS

$$\sinh x = \frac{1}{2}\left(e^x - e^{-x}\right)$$

$$\cosh x = \frac{1}{2}\left(e^x + e^{-x}\right)$$

$$\tanh x = \frac{\sinh x}{\cosh x}$$

$$\coth x = \frac{1}{\tanh x}$$

$$\csc hx = \frac{1}{\sinh x}$$

$$\sec hx = \frac{1}{\cosh x}$$

$$\sinh\left(x \pm y\right) = \sinh x \cosh y \pm \cosh x \sinh y$$

$$\cosh\left(x \pm y\right) = \cosh x \cosh y \pm \sinh x \sinh y$$

$$\tan\left(x \pm y\right) = \frac{\tan x \pm \tan y}{1 \mp \tan x \tan y}$$

B.5 DERIVATIVES

If $U = U(x)$, $V = V(x)$, and $a = $ constant,

$$\frac{d}{dx}\left(aU\right) = a\frac{dU}{dx}$$

$$\frac{d}{dx}\left(UV\right) = U\frac{dV}{dx} + V\frac{dU}{dx}$$

$$\frac{d}{dx}\left(\frac{U}{V}\right) = \frac{V\dfrac{dU}{dx} - U\dfrac{dV}{dx}}{V^2}$$

$$\frac{d}{dx}\left(aU^n\right) = naU^{n-1}$$

$$\frac{d}{dx}\left(a^U\right) = a^U \, \ln a\frac{dU}{dx}$$

$$\frac{d}{dx}\left(e^U\right) = e^U\frac{dU}{dx}$$

$$\frac{d}{dx}(\sin U) = \cos U \frac{dU}{dx}$$

$$\frac{d}{dx}(\cos U) = -\sin U \frac{dU}{dx}$$

$$\frac{d}{dx}\tan U = \frac{1}{\cos^2 U}\frac{dU}{dx}$$

B.6   INDEFINITE INTEGRALS

If $U = U(x)$, $V = V(x)$, and a = constant,

$$\int a\,dx = ax + C$$

$$\int U\,dV = UV - \int V\,dU \text{ (integration by parts)}$$

$$\int U^n dU = \frac{U^{n+1}}{n+1} + C, \quad n \neq 1$$

$$\int \frac{dU}{U} = \ln U + C$$

$$\int a^U dU = \frac{a^U}{\ln a} + C, \quad a > 0, a \neq 1$$

$$\int e^{ax} dx = \frac{1}{a}e^{ax} + C$$

$$\int x e^{ax} dx = \frac{e^{ax}}{a^2}(ax - 1) + C$$

$$\int x^2 e^{ax} dx = \frac{e^{ax}}{a^3}(a^2 x^2 - 2ax + 2) + C$$

$$\int \ln x\,dx = x\ln x - x + C$$

$$\int \sin ax\,dx = -\frac{1}{a}\cos ax + C$$

$$\int \cos ax\,dx = \frac{1}{a}\sin ax + C$$

$$\int \sin^2 ax\,dx = \frac{x}{2} - \frac{\sin 2ax}{4a} + C$$

$$\int \cos^2 ax\, dx = \frac{x}{2} + \frac{\sin 2ax}{4a} + C$$

$$\int x \sin ax\, dx = \frac{1}{a^2}\left(\sin ax - ax \cos ax\right) + C$$

$$\int x \cos ax\, dx = \frac{1}{a^2}\left(\cos ax + ax \sin ax\right) + C$$

$$\int x^2 \sin ax\, dx = \frac{1}{a^3}\left(2ax \sin ax + 2\cos ax - a^2x^2 \cos ax\right) + C$$

$$\int x^2 \cos ax\, dx = \frac{1}{a^3}\left(2ax \cos ax - 2\sin ax + a^2x^2 \sin ax\right) + C$$

$$\int e^{ax} \sin bx\, dx = \frac{e^{ax}}{a^2 + b^2}\left(a \sin bx - b \cos bx\right) + C$$

$$\int e^{ax} \cos bx\, dx = \frac{e^{ax}}{a^2 + b^2}\left(a \cos bx + b \sin bx\right) + C$$

$$\int \sin ax \sin bx\, dx = \frac{\sin(a-b)x}{2(a-b)} - \frac{\sin(a+b)x}{2(a+b)} + C, \quad a^2 \neq b^2$$

$$\int \sin ax \cos bx\, dx = -\frac{\cos(a-b)x}{2(a-b)} - \frac{\cos(a+b)x}{2(a+b)} + C, \quad a^2 \neq b^2$$

$$\int \cos ax \cos bx\, dx = \frac{\sin(a-b)x}{2(a-b)} + \frac{\sin(a+b)x}{2(a+b)} + C, \quad a^2 \neq b^2$$

$$\int \frac{dx}{a^2 + x^2} = \frac{1}{a}\tan^{-1}\frac{x}{a} + C$$

$$\int \frac{x^2 dx}{a^2 + x^2} = x - a\tan^{-1}\frac{x}{a} + C$$

$$\int \frac{dx}{\left(a^2 + x^2\right)^2} = \frac{1}{2a^2}\left(\frac{x}{x^2 + a^2} + \frac{1}{a}\tan^{-1}\frac{x}{a}\right) + C$$

B.7 DEFINITE INTEGRALS

If m and n are integers,

$$\int_0^{2\pi} \sin ax\, dx = 0$$

$$\int_0^{2\pi} \cos ax \, dx = 0$$

$$\int_0^{\pi} \sin^2 ax \, dx = \int_0^{\pi} \cos^2 ax \, dx = \frac{\pi}{2}$$

$$\int_0^{\pi} \sin mx \sin nx \, dx = \int_0^{\pi} \cos mx \cos nx \, dx = 0, \quad m \neq n$$

$$\int_0^{\pi} \sin mx \cos nx \, dx = \begin{cases} 0, & m+n = \text{even} \\ \dfrac{2m}{m^2 - n^2}, & m+n = \text{odd} \end{cases}$$

$$\int_0^{2\pi} \sin mx \sin nx \, dx = \int_{-\pi}^{\pi} \sin mx \sin nx \, dx = \begin{cases} 0, & m \neq n \\ \pi, & m \neq n \end{cases}$$

$$\int_0^{\infty} \frac{\sin ax}{x} \, dx = \begin{cases} \dfrac{\pi}{2}, & a > 0 \\ 0, & a = 0 \\ -\dfrac{\pi}{2}, & a < 0 \end{cases}$$

$$\int_0^{\infty} \frac{\sin^2 x}{x} \, dx = \frac{\pi}{2}$$

$$\int_0^{\infty} \frac{\cos bx}{x^2 + a^2} \, dx = \frac{\pi}{2a} e^{-ab}, \quad a > 0, b > 0$$

$$\int_0^{\infty} \frac{x \sin bx}{x^2 + a^2} \, dx = \frac{\pi}{2} e^{-ab}, \quad a > 0, b > 0$$

$$\int_0^{\infty} \sin cx \, dx = \int_0^{\infty} \sin c^2 x \, dx = \frac{1}{2}$$

$$\int_0^{\pi} \sin^2 nx \, dx = \int_0^{\pi} \sin^2 x \, dx = \int_0^{\pi} \cos^2 nx \, dx = \int_0^{\pi} \cos^2 x \, dx = \frac{\pi}{2}, \quad n = \text{an integer}$$

$$\int_0^{\pi} \sin mx \sin nx \, dx = \int_0^{\pi} \cos mx \cos nx \, dx = 0, \quad m \neq n, m, n \text{ integers}$$

$$\int_0^\pi \sin mx \cos nx \, dx = \begin{cases} \dfrac{2m}{m^2 - n^2}, & m+n = \text{odd} \\ 0, & m+n = \text{even} \end{cases}$$

$$\int_{-\infty}^{\infty} e^{\pm j2\pi tx} \, dx = \delta(t)$$

$$\int_0^{\infty} x^n e^{-ax} \, dx = \frac{n!}{a^{n+1}}$$

$$\int_0^{\infty} e^{-a^2 x^2} \, dx = \frac{\sqrt{\pi}}{2a}, \quad a > 0$$

$$\int_0^{\infty} x^{2n} e^{-ax^2} \, dx = \frac{1 \cdot 3 \cdot 5 \cdots (2n-1)}{2^{n+1} a^n} \sqrt{\frac{\pi}{a}}$$

$$\int_0^{\infty} x^{2n+1} e^{-ax^2} \, dx = \frac{n!}{2a^{n+1}}, \quad a > 0$$

B.8 L'HOPITAL'S RULE

If $f(0) = 0 = h(0)$, then

$$\lim_{x \to 0} \frac{f(x)}{h(x)} = \lim_{x \to 0} \frac{f'(x)}{h'(x)}$$

where the prime indicates differentiation.

B.9 SUMMATION

$$\sum_{k=1}^{N} k = \frac{1}{2} N(N+1)$$

$$\sum_{k=1}^{N} k^2 = \frac{1}{6} N(N+1)(2N+1)$$

$$\sum_{k=1}^{N} k^3 = \frac{1}{4} N^2 (N+1)^2$$

$$\sum_{k=0}^{N} a^k = \frac{a^{N+1} - 1}{a - 1} \quad a \neq 1$$

$$\sum_{k=M}^{N} a^k = \frac{a^{N+1}-a^M}{a-1} \quad a \neq 1$$

$$\sum_{k=0}^{N} \binom{N}{k} a^{N-k} b^k = (a+b)^N, \quad \text{where} \quad \binom{N}{k} = \frac{N!}{(N-k)!k!}$$

B.10   NUMERICAL INTEGRATION APPROXIMATIONS

$$\text{TRAP}(n) = \frac{b-a}{2n}\Big[f(x_0)+2f(x_1)+2f(x_2)+\cdots+2f(x_{n-1})+f(x_n)\Big]$$

$$\text{SIMP}(n) = \frac{b-a}{3n}\Big[f(x_0)+4f(x_1)+2f(x_2)+4f(x_3)+\cdots+2f(x_{n-2})+4f(x_{n-1})+f(x_n)\Big]$$

B.11   POWERS OF THE TRIG FUNCTIONS

1) Integrals of the form: $\int \sin^m x \cos^n x \, dx$ m or n odd

Strategy: If m is odd, save a sine factor and convert to cosine
If n is odd, save a cosine factor and convert to sine

2) Integrals of the form: $\int \sin^m x \cos^n x \, dx$ m and n even and non-negative

Strategy: use ½ angle identities:
$\cos^2 x = \dfrac{1+\cos(2x)}{2}$; $\sin^2 x = \dfrac{1-\cos(2x)}{2}$

3) Integrals of the form: $\int \sec^m x \tan^n x \, dx$; if m is even

Strategy: Save a $\sec^2 x$ and convert to a tangent

4) Integrals of the form: $\int \sec^m x \tan^n x \, dx$; if n is odd

Strategy: Save a $\sec x \tan x$ and convert to secant

5) $\int \tan^n x \, dx$ n any positive integer

Strategy: convert a $\tan^2 x$ to $\sec^2 x - 1$ and distribute; repeat if necessary

6) $\int \sec^m x \, dx$; m is odd

Strategy: Integrate by parts

Appendix C: MATLAB

The word of MATLAB is an abbreviation for Matrix Laboratory implying that MATLAB is a MATLAB is available for Macintosh, Unix, and Windows operating systems.

The appendix introduces the reader to programming with the software package MATLAB. It is assumed that the reader has had some previous experience with a high-level programming language and is familiar with the techniques of writing loops, branching using logical relations, calling subroutines, and editing; therefore, this appendix starts with the essential and present computational tools that employ matrices and vectors/arrays to carry out numerical analysis, signal processing, and scientific visualization tasks.

A brief introduction to MATLAB is presented in this appendix and Chapter 1 are enough to be able to solve problems in this book. Other information on MATLAB required in this book is provided on a chapter-to-chapter basis as needed.

MATLAB has become a powerful tool for technical professionals worldwide. A student version of MATLAB is available for PCs. A copy of MATLAB can be obtained from:

The Mathworks, Inc.
3 Apple Hill Drive
Natick, MA 01760-2098
Phone:(508) 647-7000
Website: http://www.mathworks.com
To begin to use MATLAB, we use the following operators. Type commands into MATLAB prompt ">>" in the command window. To get help, type
>> help

C.1 MATLAB BASICS

To find additional information about commands, options, and examples, the reader is urged to make use of the online help facility and the Reference and User's Guides that accompany the software.

TABLE C.1 BASIC OPERATIONS

Operation	MATLAB formula
Addition	a + b
Subtraction	a − b
Division (right)	a/b (means $a \div b$)
Division (left)	a\b (means $b \div a$)
Multiplication	a*b
Power	a^b

TABLE C.2 BASIC FUNCTIONS

Function	Remark
abs(x)	Absolute value or complex magnitude of x
acos, acosh(x)	Inverse cosine and inverse hyperbolic cosine of x in radians
acot, acoth(x)	Inverse cotangent and inverse hyperbolic cotangent of x in radians
angle(x)	Phase angle (in radian) of a complex number x
asin, asinh(x)	Inverse sine and inverse hyperbolic sine of x in radians
atan, atanh(x)	Inverse tangent and inverse hyperbolic tangent of x in radians
conj(x)	Complex conjugate of x
cos, cosh(x)	Cosine and hyperbolic cosine of x in radian
cot, coth(x)	Cotangent and hyperbolic cotangent of x in radian
exp(x)	Exponential of x
fix	Round toward zero
imag(x)	Imaginary part of a complex number x
log(x)	Natural logarithm of x
log2(x)	Logarithm of x to base 2
log10(x)	Common logarithms (base 10) of x
real(x)	Real part of a complex number x
sin, sinh(x)	Sine and hyperbolic sine of x in radian
sqrt (x)	Square root of x
tan, tanh(x)	Tangent and hyperbolic tangent of x in radian

TABLE C.3 MATRIX OPERATIONS

Operation	Remark
A'	Finds the transpose of matrix A
det(A)	Evaluates the determinant of matrix A
inv(A)	Calculates the inverse of matrix A
eig(A)	Determines the eigenvalues of matrix A
diag(A)	Finds the diagonal elements of matrix A
expm(A)	Exponential of matrix A

TABLE C.4 SPECIAL MATRICES, VARIABLES, AND CONSTANTS

Matrix/Variable/Constant	Remark
eye	Identity matrix
ones	An array of ones
zeros	An array of zeros
i or j	Imaginary unit or sqrt(-1)
pi	3.142
NaN	Not a number
inf	Infinity
eps	A very small number, 2.2e$-$16
rand	Random element

TABLE C.5 VECTOR OPERATIONS

Vector	Remark
sum(a)	Sum of vector elements
mean(a)	Mean of vector elements
std(a)	Standard deviation
max(a)	Maximum
min(a)	Minimum

TABLE C.6 COMPLEX NUMBERS

real(A)	The real part of A
imag(A)	The imaginary part of A
conj(A)	The complex conjugate of A
abs(A)	The modulus of A
angle(A)	The phase angle of A

TABLE C.7 MATRIX OPERATIONS

Operation	MATLAB formula
Matrix Addition	+
Matrix Subtraction	−
Matrix Multiplication	*
Right Matrix Division	/
Left Matrix Division	\
Raise to a power	^
Transpose matrix	'

TABLE C.8 ARRAY OPERATIONS

Operation	MATLAB formula
Array Multiplication	.*
Right Array Division	./
Left Array Division	.\
Raise to a power	.^

TABLE C.9 MATRICES AND FUNCTIONS

Operation	Remark
sqrtm(x)	The matrix square root
expm(x)	The matrix exponential base e
logm(x)	The matrix natural logarithm

TABLE C.10 UTILITY MATRICES

Operation	Remark
zeros(n)	$n*n$ matrix where each element is 0
zeros(m,n)	$m*n$ matrix where each element is 0
ones(n)	$n*n$ matrix where each element is 1
ones(m,n)	$m*n$ matrix where each element is 1
rand(n)	$n*n$ matrix of random numbers
rand(m,n)	$m*n$ matrix of random numbers
eye(n)	$n*n$ identity matrix

TABLE C.11 RELATIONAL AND LOGICAL OPERATIONS

Operation	MATLAB formula
Less than	<
Less than or equal	<=
Greater than	>
Greater than or equal	>=
Equal	==
Not equal	~=

TABLE C.12 BOOLEAN LOGIC OPERATIONS

Operation	MATLAB formula
AND	&
OR		
NOT	~

TABLE C.13 VARIABLE CONTROL COMMANDS

Operation	Remark
who	List all the variables in memory
whos	List all the variables in memory with more information
clear	Remove all variables from memory
clear <variable>	Remove specified variables from memory

TABLE C.14 FILE CONTROL COMMANDS

Operation	Remark
dir	List the contents of the current director
ls	List the contents of the current directory
what	List the MATLAB files in the current directory
cd <directory>	Change the current directory
type <filename>	Display the contents of a text or.m file
delete <filename>	Delete a file
diary <filename>	Record all commands and results to a file
Diary	off Stop above

TABLE C.15 SAVING, EXPORTING AND IMPORTING DATA

Operation	Remark
Save	Save all variable to the file MATLAB.mat
load	Load in variables from the file MATLAB.mat
save \<filename>	Save all the variable to the file filename.mat
load \<filename>	Load in the variables from the file filename.mat
save \<filename> \<variable>	Save only the variable variable to the file filename.mat
load \<filename> \<variable>	Load in only the variable variable from the file filename.mat
save \<filename> \<variable> -ascii	Save the variable to the text file filename
load \<filename>.\<ext>	Load from the text file, to the variable called filename
save \<filename> \<variable> -ascii -	double Save variable to the text file filename using double precision
csvwrite(\<filename>.cvs, M)	Write matrix M to csv
file. M=csvread(\<filename>.csv)	Load from the csv file, into variable M

C.2 USING MATLAB TO PLOT

To plot using MATLAB is easy. For a two-dimensional plot, use the plot command with *two* arguments, such as

>> plot(*xdata*,*ydata*)

where *xdata* and *ydata* are vectors of the same length containing the data to be plotted.

MATLAB will let you graph multiple plots together and distinguish them with different colors.

It is obtained with the command plot (*xdata*, *ydata*, "color"), where the color is indicated by using a character string from the options listed in Table C.5.

TABLE C.16 VARIOUS COLOR AND LINE TYPES

y	yellow	.	point
m	magenta	o	circle
c	cyan	x	x-mark
r	red	+	plus
g	green	–	solid
b	blue	*	star
w	white	:	dotted
k	black	-.	dashdot
		--	dashed

TABLE C.17 PLOTTING COMMANDS

Command	Comments
bar(x,y)	a bar graph
contour(z)	a contour plot
errorbar(x,y,l,u)	a plot with error bars
hist(x)	a histogram of the data
plot3(x,y,z)	a three-dimensional version of plot()
polar(r, angle)	a polar coordinate plot
stairs(x,y)	a stairstep plot
stem(n,x)	plots the data sequence as stems
subplot(m,n,p)	multiple (m-by-n) plots per window
surf(x,y,x,c)	a plot of a three-dimensional colored surface
hold on	Hold the plot on the screen
hold off	Release the plot on the screen
hold	Toggle the hold state

Index

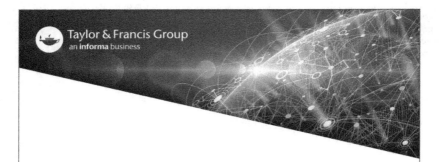